The Soviet Biological Weapons Program and Its Legacy in Today's Russia

Raymond A. Zilinskas

Center for the Study of Weapons of Mass Destruction
Occasional Paper, No. 11

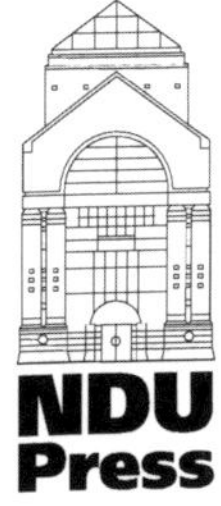

National Defense University Press
Washington, D.C.
July 2016

First printing, July 2016

Contents

Acknowledgments

Several individuals contributed greatly to the success of this project, and the author wishes to thank them for their encouragement and assistance. In particular, I am indebted to Dr. Mikhail Tsypkin, an associate professor in the Department of National Security Affairs at the Naval Postgraduate School and an expert on Soviet military affairs, who reviewed a draft of this paper and helped me navigate the *labyrinthine* Soviet bureaucracy and its difficult decisionmaking process. I also thank Dr. Seth Carus and Mr. Paul Bernstein for their guidance while the occasional paper was being developed and Ms. Helen Zilinskas for having edited a late version of the draft paper.

Executive Summary

In its first Weapons of Mass Destruction (WMD) Case Study, the Center for the Study of Weapons of Mass Destruction (CSWMD) at the National Defense University examined President Richard M. Nixon's decision, on November 25, 1969, to terminate the U.S. offensive biological weapons program.[1] This occasional paper seeks to explain why the Soviet government, at approximately the same time, decided to do essentially the opposite, namely, to establish a large biological warfare (BW) program that would be driven by newly discovered and powerful biotechnologies. By introducing the innovation of recombinant DNA technology—commonly referred to as genetic engineering—the Soviets were attempting to create bacterial and viral strains that were more useful for military purposes than were strains found in nature.

Nixon's decision was widely publicized and documents revealing the reasons behind it are, in the main, available to the public in the National Archives, the Nixon Presidential Library in San Clemente, California, and publications written by members of the Nixon administration. In sharp contrast, the Soviet decision was highly secret at the time and remains so to this day. All contemporary documents pertaining to the Soviet offensive BW program remain classified and none of the military officers and officials who operated it has spoken or written about it except to deny that it existed or to belie its offensive intent.

The information that has become available about the program has been divulged by scientists and administrators who previously worked in the civilian component of the Soviet BW program, called Biopreparat. Some of them had defected to Western countries and there told their stories, while others chose to remain in Russia after the Soviet Union dissolved in December 1991 and later divulged details of their past secret activities in their own publications, or in broadcasts or print interviews. However, Biopreparat employees, even those who held managerial positions, did not have sufficiently high clearances to be informed about high-level BW-related decisionmaking. Decisions such as those that instituted what in effect was a new BW program, and ordered the Soviet Ministry of Defense (MOD) to develop strategies and tactics for the use of biological weapons, were made at the highest levels by members of the Politburo and Central Committee of the Communist Party (CCCP) and the MOD's General Staff (GS).

Nevertheless, some information pertaining to the establishment of Biopreparat, the planning of programs to research and develop weapons against humans (codenamed *Ferment*) and animals and plants (codenamed *Ekology*), and the accomplishments of these program have become known because Biopreparat scientists learned about them from military scientists who divulged some of this knowledge while working together or in relaxed situations. Thus, the two authors of an extensive history of the Soviet BW program, one of whom is the author of this paper, were able to collect sufficient information from their interviews with Biopreparat employees, autobiographies written by weapons scientists, and articles written by investigative Russian reporters to describe and discuss important aspects of Soviet decisionmaking concerning BW.[2] While this paper draws largely on the contents of this book, additional information comes from sources listed in the endnotes, particularly from the studies on Soviet military decisionmaking conducted by John G. Hines, Ellis M. Mishulovich, and John F. Shull.[3]

In historical terms, the Soviet BW program had two so-called "generations," defined as distinct periods of time during which types of weapons were developed from earlier types.[4] The first generation of the Soviet BW program commenced about 1928 and was based on naturally occurring pathogens that had caused devastating epidemics during World War I and the subsequent Russian Civil War. The second generation began approximately in 1972 when the decision was made at the highest political level to institute a research and development (R&D) system that utilized newly discovered techniques of genetic engineering to create novel or enhanced bacterial and viral strains that were better adapted for BW purposes than strains found in nature. President Boris Yeltsin ordered the cessation of the offensive BW program some months after the Soviet Union dissolved in December 1991 and in 1992 publically stated that it had conducted an offensive BW program in violation of the Biological and Toxin Weapons Convention. However, after Vladimir Putin was elected president, high-level Russian officials have lied about the Soviet BW program, stating that it was strictly a defensive program that had not broken international law. As is discussed later in this paper, elements of the Soviet offensive BW program continue in Russia and may provide the basis for a third-generation BW program supported by the current leadership.

The first section of this paper describes the Soviet BW program's first generation, including its establishment, work plan and operations, and accomplishments. The second section focuses on "establishing the conditions" for the Soviet decision that was made sometime during 1969–1971 to establish and operate the second-generation BW program. Conditions that are considered include the geopolitical challenges as perceived by the Soviet government, the decisionmaking process for military acquisitions, and the inferior state of the biosciences in the Soviet Union at that time, which stimulated Soviet bioscientists to "play the military card" in order to introduce genetic engineering into the Union of Soviet Socialist Republics' (USSR's) bioscience establishment. The final section has two sub-sections. The first summarizes the key factors that drove Soviet decisionmaking in the early 1970s to institute a huge offensive BW program. The second informs readers that even before Vladimir Putin was elected president for the second time, he openly stated that new weapons were to be developed using high technologies including "genetics." Based on this promise, and considering the secrecy that still keeps the military biological institutes and anti-plague institutes closed to outsiders, the paper discusses the possibility that the Putin administration may institute a third-generation BW program. The appendix consists of a short biography of the Soviet general Yefim Ivanovich Smirnov who was for many years in charge of the Soviet BW program.

The Soviet Union's First-Generation Biological Warfare Program, 1926–1972

Russian armies suffered heavy losses from disease during all three of the major conflicts they were involved in at the beginning of the 20th century: the Russo-Japanese War (1904–1905), World War I (1914–1917), and the civil war between the Red and White forces (1918–1921). Disease caused more casualties in all of these conflicts than did weapons.[5] Armies and civilians were especially devastated by typhus.[6] A Soviet epidemiologist writes, "There were 20 to 30 million cases of typhus between 1918 and 1922 in the territories controlled by the new Soviet Republic, and a mortality rate of around 10 percent."[7] Vladimir I. Lenin is quoted as having stated, "We are suffering from a desperate crisis. . . . A scourge is assailing us, lice, and the typhus that is mowing down our troops. Either the lice will defeat socialism, or socialism will defeat the lice!"[8]

Biological warfare was not waged on the Eastern Front during World War I,[9] but with Russian troops having experienced the full effects of German chemical weapons,[10] the Bolshevik government that took power after the 1917 revolution was intent on creating a chemical industry that could serve both civilian and military purposes. To integrate chemical weapons into its force structure, the Worker's and Peasant's Red Army (RKKA) created the Military Chemical Agency in 1925 under the directorship of Yakov M. Fishman, who was to remain in this position until 1937 when he fell victim to Joseph Stalin's Great Purge. One of Fishman's first acts as director was to set up a small BW laboratory in Moscow, which was named the Scientific Research Institute of Health and was headed by Nikolay N. Ginsburg. In 1928, Fishman submitted a four-part progress report to Kliment Y. Voroshilov, People's Commissar for Military and Naval Affairs. The first part described the work that had been done by Ginsburg (see below), which demonstrated the feasibility of BW. The second assessed the potential uses of bacteria for warfare and sabotage, including as payloads in artillery shells and bombs. The third part presented a plan for the organization of military biology, and the fourth presented a plan for organizing defenses against biological attacks. Acting on Fishman's recommendations, the Revolutionary Military Council in 1928 issued a secret decree that ordered the establishment of offensive and defensive

BW programs and designated the Military Chemical Agency to manage both.[11] In addition, a civilian agency, the People's Health Commissariat, was ordered to coordinate and execute military tasks related to BW.[12] At that time, the People's Health Commissariat was operating a substantial network consisting of at least 35 institutions working in such disciplines as epidemiology, genetics, immunology, microbiology, virology, and plague protection.

Opinions differ as to how the decree was implemented. The findings of research conducted by Bojtzov and Geissler are the most trustworthy.[13] Their research revealed that the BW program was headed by Ginsburg and initially focused on weaponizing *Bacillus anthracis*[14] and *Clostridium botulinum*[15] and developing efficient methods of disinfecting persons and equipment contaminated with pathogens. According to Fishman's 1928 progress report, Ginsburg's group attempted to enhance properties of *B. anthracis* spores, which already in their natural state are well suited for purposes of BW, being both virulent and hardy, but which conceivably could be made even deadlier by using classical microbiology techniques of mutation, selection, and propagation. Military scientists would expose natural pathogen strains to mutation-inducing chemicals or irradiation, then would recover mutants that possessed improved or enhanced characteristics related to infectivity, virulence, and hardiness, and would propagate promising mutants in mass. In this research, various types of animals were used as test subjects, including cats, rabbits, goats, and horses. The candidate BW agents either were injected into test animals or dispersed as aerosols in closed chambers that contained animals. Typically, test animals died within 2 or 3 days of exposure. Another dispersal method used explosives to spread a quantity of the BW agents. In this case, the explosion created an aerosol whose particles contained the microorganisms. Ginsburg's laboratory also studied *Vibrio cholerae*[16] and *Yersinia pestis*[17] to ascertain whether they would be useful as BW agents; the latter eventually was weaponized.

In 1933, the RKKA established the Vaccine-Serum Laboratory in the village of Vlasikha, outside Moscow. Its objective was to develop and manufacture vaccines and sera against common infectious diseases. Professor Ivan M. Velikanov, then the head of the M.V. Lomonosov Moscow State University (MSU) microbiology department, was appointed director of the laboratory. Also in 1933, the

United State Political Administration (OGPU) set up a laboratory, named the Special Purpose Bureau, to research highly infectious diseases,[18] located on the property of the former Pokrovsky Monastery in the small town of Suzdal in the Vladimir *oblast*.[19] The facility's staff members lived in cells formerly occupied by monks and were not allowed to leave the grounds without special permission. The monastery's chapel served as an animal facility, containing cages in which marmosets, guinea pigs, and rats were kept. In addition, sheep and two camels used as test subjects grazed in the monastery's yard. Bureau scientists studied pathogens that cause cholera, plague, tetanus, and malaria.

In late 1933, the Vaccine-Serum Laboratory and the Special Purpose Bureau were combined to create the RKKA Military Medical Scientific Institute, which continued to be headquartered at Vlasikha. In 1934, the institute was renamed the RKKA Biotechnical Institute. An accident in 1937 led to its relocation to Gorodomlya Island, located on Lake Seliger in the Kaliningrad oblast,[20] about 350 kilometers northwest of Moscow. The reason for the move was that the deputy director of the institute, Abram L. Berlin, unknowingly infected himself with *Y. pestis* during an experiment to develop a new anti-plague vaccine. After being infected, but before he showed symptoms, Berlin traveled to Moscow to report on the vaccine's progress. While there, he infected two other people, and all three died of bubonic plague. Fortunately, the local health authorities acted quickly and effectively, thereby preventing the disease from spreading.[21] However, Kremlin officials concluded that the institute had endangered Moscow's population and therefore ordered it to be relocated far away from the city. Berlin was the first known Soviet BW research scientist to be killed by a pathogen under study.

The RKKA Biotechnical Institute was renamed the Medical-Technical Institute of the RKKA (STI) in 1940. Soon after the German invasion that began in June 1941, Soviet authorities feared that the Kaliningrad oblast would be overrun by German forces, and ordered the institute relocated to Saratov; it was renamed the Scientific Research Institute of Epidemiology and Hygiene. As the Battle of Stalingrad raged from August 23, 1942 to February 2, 1943, the Luftwaffe mounted air attacks on nearby cities, including Saratov. To safeguard the institute, it was moved to Kirov City, where it was headquartered in an old hospital, and where it remains to this day.[22] For convenience in this paper, it is henceforth called the Kirov Institute.

Many able microbiologists fell victim to Stalin's purges in the late 1930s, including Fishman, the founder of the Soviet Union's BW program.[23] Of more importance to this study, the People's Commissariat for Internal Affairs (NKVD) arrested many microbiologists to secure expertise that it could exploit at small cost. As has been vividly described by Aleksandr I. Solzhenitsyn, the NKVD often placed imprisoned scientists and engineers in groups called *sharaga* and allowed them to work much as they did in their "free" days.[24] For example, N.A. Gaysky, a specialist on *Francisella tularensis,* was ordered to work as a member of a sharaga developing a vaccine against tularemia at the Third Experimental Laboratory of the Red Army.[25] Similarly, an expert on rickettsiae, P.F. Zdrodovsky, worked in a sharaga while imprisoned, as did L.A. Zilber, who had proposed that viruses are the cause of some cancers. After their release, some of these scientists continued to work willingly at the institutions where they had been imprisoned. Despite the decimation of Soviet scientists by Stalin, presumably a sufficient number of them survived for the Soviet Union to have maintained an active BW program until World War II and beyond.

On February 22, 1938, the world learned that the Soviet Union possessed both biological and chemical weapons. In a speech reported by Western media, Marshal of the Soviet Union Kliment Y. Voroshilov stated:

> *Ten years ago or more the Soviet Union signed a convention abolishing the use of poison gas and bacteriological warfare.*[26] *To that we still adhere, but if our enemies use such methods against us, I tell you that we are prepared—fully prepared—to use them also and to use them against aggressors on their own soil.*[27]

The civilian leaders in the Kremlin considered Voroshilov's remark to have been a gross indiscretion. If he had stated that the USSR "reserves the right" under the Geneva Protocol to reply in kind, he probably would have had no problems with his superiors, but by asserting that "we are prepared—fully prepared—to use them," he in effect was telling the world that the Soviets possessed both chemical and bacteriological weapons. This was contrary to the image that the Soviet Union was trying to project: that only ruthless capitalist states possessed these weapons while the Soviets only sought to defend against them.

In the interwar years, Western intelligence agencies knew very little about the Soviet BW establishment. However, German forces that invaded the Soviet Union in 1941 almost immediately captured several hundred thousand prisoners of war (POWs), some of whom in peacetime had worked in various military facilities. German intelligence agencies set up units to interview POWs who could be expected to have knowledge of various Soviet activities. Two Wehrmacht intelligence officers, Walter Hirsch and Heinrich Kliewe, specialized in gathering information on Soviet chemical and biological warfare programs. Both were captured by American forces in 1945, and then willingly provided huge amounts of information that had been collected from Soviet POWs. Some of this information proved to be erroneous, but much was worthwhile. According to information gathered by Kliewe,[28] by the time World War II broke out, three institutes were involved in offensive BW activities in the Moscow oblast: Ginsburg's institute, the Moscow Chemical-Pharmaceutical Institute, and the Saratov Institute for Microbiology and Epidemiology. In the Leningrad oblast, four institutes were supposedly involved in BW research and development: the Zlatogorov-Maslokovich Laboratory at the Leningrad Veterinary and Zoological Technical Institute, the Bacteriological Institute of Leningrad, an unnamed facility at the Kronstadt naval base,[29] and an unnamed research station on the shore of Lake Ladoga. These institutions largely focused their efforts to weaponize *B. anthracis* and *Y. pestis*, although they did some work to develop BW agents against cattle, including the virus that causes foot-and-mouth disease.

The Soviets established three open-air test sites before World War II. The first, in 1925, was at Tomka (renamed Staryye Shikhany, in 1933), near Volsk on the Volga River. Called the Central Chemical Proving Ground (*Tsentralny Khimichesky Poligon*, or *TsKhP*), or more simply the Volsk Polygon, it covered approximately 100 square kilometers. During the time of the German-USSR accord in the 1920s, military units from both sides trained there together, and conducted exercises involving the use of both conventional and chemical weapons. The other two open-air test sites were located on islands. Gorodomlya Island had a 10-square-kilometer site on which weapons containing pathogens causing foot-and-mouth disease, leprosy, plague, and tularemia were tested.[30] The second island, which was to become the favored site for large-scale open-air testing of biological agents and weapons, as well as defensive equipment and measures, was Vozrozhdeniye Island in the Aral Sea. The facility there, code-named Aralsk-7, became fully operational in 1936, but was closed down during World War II. It was reopened in the 1950s, after which it was the only open-air test site in the Soviet Union for the realistic testing of weapons armed with all the different types of pathogens weaponized during both the first and second generations of the Soviet BW program. The extreme isolation of Aralsk-7 kept its activities far away from prying eyes and afforded a high level of biosafety for open-air tests.

For the sake of comparison, by the time World War II commenced, Japan was the only major nation with an offensive BW program approximating the size and status of the Soviet Union's.[31] The major Japanese military unit dedicated to developing biological weapons, Unit 731, was headquartered at Ping Fan in Manchuria, only a few hundred kilometers from the Soviet border. France's small program, which was active in the 1930s, was terminated when German armies were close to occupying that country in 1940.[32] The United Kingdom (UK) had started a BW program in 1937, but it did not reach full maturity until the early 1940s and never reached anything near the size of the Japanese and Soviet programs.[33] Canada had also begun considering BW in the late 1930s and, in cooperation with the UK and the United States, was to have a full-scale program by the mid-1940s.[34] The United States began to consider establishing a BW program in 1942 and did so in 1943.[35] Germany,[36] Italy,[37] and Poland[38] had no offensive BW programs and, at most, rudimentary defensive programs.

In 1939, Stalin placed his Minister of Internal Affairs, Lavrenty P. Beria, in overall command of the Soviet BW program. In practice, the Main Military Medical Directorate of the Red Army, headed by Colonel-General Yefim I. Smirnov,[39] had responsibility for its day-to-day operations. Smirnov, described in a Russian publication as being "a distinguished organizer and theorist of military and civilian health,"[40] was at that time a rising star in the military medical establishment, and from the late 1940s through the early 1980s was one of the main planners and proponents, perhaps *the* main one, for the development of biological weapons (see appendix for a biography of Smirnov).[41]

The Soviet Union had not only an extensive offensive BW program, but also a substantial program to defend against both biological weapons and natural infectious diseases. The offensive and defensive BW programs were conducted side by side, often in the same institutions. The Soviet defensive research program had seven objectives:

- to develop and improve vaccines against BW agents that enemies might use
- to develop methods and protocols for immunization utilizing vaccines and other protective substances
- to develop protocols for the emergency treatment of soldiers exposed to BW agents, including diagnosis
- to develop methods, means, and regimes for disinfecting personnel and equipment contaminated by BW agents
- to develop methods for identifying BW agents and clarifying indications of biological attacks
- to develop and test field detection systems for BW agents
- to assess the possible damage of the various "recipes"[42] that an enemy might employ against the Soviet Union.[43]

It is not known exactly when these objectives were formulated, but they continued to guide such research until the USSR dissolved in 1991.

The major defense efforts in the 1930s and leading up to World War II at the Kirov Institute and its predecessors sought to develop live vaccines against anthrax, plague, tularemia, brucellosis, and tuberculosis.[44] However, the highest priority was to develop an efficacious anthrax vaccine. The major R&D to this end was conducted by military scientists at the Research Institute of Epidemiology and Hygiene starting in 1935. Orlov wrote that this vaccine R&D was deemed so important that Smirnov was personally required to report on its progress to Beria and Stalin.[45] By 1940, Soviet microbiologists had developed two avirulent strains of *B. anthracis*, named STI-1[46] and No. 3, which were derived from virulent parent strains. When used as a trial vaccine in animals, the STI-1 strain protected 60 percent of guinea pigs, 70 percent of rabbits, and 97 percent of sheep that had been deliberately infected by virulent strains of *B. anthracis*. Based on these good results, more than 2 million domestic animals were vaccinated with the STI-1 strain of vaccine during World War II. After the war, many more animals were vaccinated—38.4 million in 1947, increasing to 140 million in 1960. As a result, the number of domesticated animals that died from anthrax in the USSR decreased from 30,500 in 1947 to just 3,500 in 1960.[47]

The same Kirov Institute scientists who had developed the animal anthrax vaccine developed a similar vaccine for use in humans. The vaccine proved safe when administered to volunteers in May 1943 using the technique of scarification.[48] In 1944, with the Red Army poised to liberate Rumania, Soviet military epidemic intelligence determined that there was a substantial threat of anthrax in that country that might affect not only animals, but humans. Accordingly, 9,000 men from the units assigned to invade Rumania were vaccinated against anthrax.[49] Orlov asserts that none of these contracted the disease. (Orlov does not say how many of the non-vaccinated troops were stricken with anthrax.) The Ministry of Health licensed the scarification vaccine for general use against anthrax in 1953 and an improved injectable vaccine in 1959.[50] Other vaccines were also developed by military scientists, such as those against plague and tularemia.

From the activities discussed in the preceding paragraphs, it appears that during World War II the Soviets concentrated more on developing defenses against natural infectious diseases that threatened its soldiers and draft animals than on their offensive BW program. This situation changed as the Soviet offensive BW program was given a boost in the late 1940s for three reasons. First, Soviet leaders learned of the large, brutish Japanese program. After the Red Army invaded Manchuria on August 8, 1945, and moved quickly toward the Pacific Ocean, it overran Ping Fan where Unit 731 was headquartered. As they advanced, Red Army troops captured scientists and medical doctors who had staffed the unit's laboratories and test facilities.[51] Interrogations revealed the appalling record of the Japanese BW program,[52] the extent and sophistication of which likely gave Soviet leaders an indication of how powerful biological weapons could be and provided them with additional knowledge and know-how. The Soviets also learned, probably for the first time, that in the late 1930s, on "orders of the Japanese Kwantung Army, Detachment 100 had systematically sent bacteriological parties to the borders of the Soviet Union, where they contaminated water sources."[53] There is no information on the success or failure of these attempts at biological sabotage. Perhaps of more consequence, the Soviets were informed of the Kan-Toku-En Plan for the Japanese attack on the Soviet Union, which was approved in 1941 but never implemented. This plan included the deployment of units that would disseminate plague-infested fleas against the Red Army and conduct sabotage behind its positions.[54]

Second, the influential American scientist Theodor Rosebury wrote an article and book shortly after World War II that had the unanticipated effect of convincing Soviet civilian and military leaders of the utility of biological weapons.[55] Some of the BW scientists interviewed by Leitenberg and Zilinskas stated that Rosebury's article and book were the main determinant of the Soviet government's decision to bolster its BW program in the 1950s.[56] Ten years later, a former U.S. general published two works that served to reinforce Soviet officials' belief in the power of biological weapons.[57]

Third, soon after World War II ended, the Soviet government learned of the joint BW programs of the United States, UK, and Canada.[58] Soviet authors described in great detail the large size of these programs, the intent of these states to use biological weapons in tandem with nuclear and chemical weapons, and the perversity of the United States not having joined the Geneva Protocol (which prohibits the use of bacteriological weapons).[59] Soviet leadership might have been motivated to build a BW program in order to match these capabilities.

In 1947, the Main Military Medical Directorate of the Red Army was renamed the Main Military Medical Directorate of the Armed Forces of the USSR, with Y.I. Smirnov as its head.[60] However, soon thereafter he was promoted to Minister of Health, a post he kept until December 1952.[61] At that time, for unknown reasons, Stalin lost confidence in Smirnov and he was demoted to a low staff position. But sometime after Stalin's death, on March 5, 1953, the USSR Council of Ministers transferred the responsibilities of the Main Military Medical Directorate to the MOD's 7th Directorate of the General Staff, and in August 1953 appointed Y.I. Smirnov as its head. About the same time, the Kirov Institute was designated as the lead agency for all Soviet BW-related R&D.

Orlov writes that, having recognized the growing threat of BW, the Soviet leadership accelerated "development of means to protect the population and the army against biological weapons."[62] It appointed the Marshal of the Soviet Union, Ivan Kh. Bagramyan, as head of the domestic defensive program and Y.I. Smirnov and General Piotr N. Burgasov as his deputies.[63]

As part of capability building, the MOD decided to construct and equip two new research units. The first was an institute dedicated to the study and weaponization of viruses and *Rickettsia*. This came about after an internal assessment concluded that the Soviet army's need for a "bacterial component" was "covered," but not the "virological component."[64] The assessment noted that the country "had only a single recently organized [within the past 5 years] civilian virology institute of the USSR Academy of Medical Sciences." Further, "for a number of reasons, the latter [civilian virology institute] naturally could not engage in assessing the threat of viruses being used for military purposes."[65] The new virology institute was established near Zagorsk (now Sergiyev Posad); henceforth it is called Zagorsk Institute in this paper.

The second new research unit was an institute whose stated purpose was to conduct research on military hygiene. The MOD took over the former Cherkassy-Sverdlovsk Infantry School in Sverdlovsk (now called Yekaterinburg) and rebuilt it, so in actuality it housed development laboratories and biological production equipment. The first group of scientists and technicians to staff the new plant began working in 1949. In 1960, this branch was separated from the Kirov Institute and renamed the Military Technical Scientific Research Institute (henceforth, Sverdlovsk Institute).[66] The institute was located within Military Compound 19 and its major function was to mass-produce pathogens used to arm biological weapons. In 1979, it became infamous when an accident at one of its production facilities led to the escape of a large number of *Bacillus anthracis* spores that were carried by prevailing winds over parts of Sverdlovsk and then to six villages outside the city. Shortly thereafter, the production plant was closed and its production function was moved to Stepnogorsk in Kazakhstan.[67]

In addition to MOD institutes dedicated to biological weapons and biological defense, the ministries of Agriculture, Internal Affairs, and Health each operated BW-related R&D units.[68] Further, both the USSR Academy of Sciences (USSR-AS) and the USSR Academy of Medical Sciences provided expert assistance to the offensive and defensive BW programs. Next to nothing is known about the activities or involvement of these agencies with BW during the pre-1970 era.

The Soviet BW program appears to have been energized in the period immediately following World War II mainly by the infusion of practical information gained from the Japanese BW experience, including the reopening and substantial build-up of Aralsk-7. As a result, a number of pathogens were weaponized, including *B. anthracis*, *F. tularensis*, *Y. pestis*, *Coxiella burnetii*, *Brucella suis*, Venezuelan equine encephalitis virus, and botulinum neurotoxin (see table 1). Perhaps the most powerful of the biological weapons generated at this time by the first-generation BW program was one based on the variola virus (the cause of smallpox). The first-generation BW program experienced a second accident at the Aralsk-7 test site in 1972, when variola virus escaped during an open-air test of biological weapons and was carried by the prevailing wind to a research ship where a scientist was infected. When the ship reached its homeport in Aralsk, the infected person had spread the virus to 11 others, 5 of whom died.[69]

For unknown reasons, the Soviet BW program seemed to be treading water in the late 1960s, with no breakthroughs or significant advances. Biopreparat scientist Igor Domaradskij claimed that the program in this period was conducted in a desultory way and was so unproductive that the Soviet military command considered terminating it.[70] This assessment was supported by former intelligence analyst and Soviet specialist Raymond Garthoff, who noted:

> *On August 17, 1967, a top secret joint decree issued by the Central Committee-Council of Ministers of the Soviet Union reviewed the evidence for what was seen as an extensive and successful U.S.*

Table 1. Soviet First-Generation Biological Warfare Agents

Bacteria	Disease caused by
Bacillus anthracis	Inhalation anthrax
Brucella melitensis (Brucella suis?)	Brucellosis (undulant fever)
Coxiella burnetii	Q fever
Francisella tularensis	Tularemia
Rickettsia prowazekii	Epidemic typhus
Yersinia pestis	Pneumonic plague
Virus	
Variola major	Smallpox
Venezuelan equine encephalomyelitis virus	Encephalitis
Toxin	
Botulinum neurotoxin	Botulism

Source: Milton Leitenberg and Raymond A. Zilinskas, *The Soviet Biological Weapons Program: A History* (Cambridge, MA: Harvard University Press, 2012), 46.

> *program in the field of chemical and biological warfare* [CBW]. *The decree called for corresponding Soviet CBW preparations. Although my attempts to obtain this decree in the Russian archives have so far been unsuccessful, I was able to track down a reference to it in the index of the still-closed files of the Central Committee.*[71]

There are two probable explanations for why the Soviet BW program faltered. First, most of the MOD's highest officials came to believe that nuclear weapons far superseded other capabilities, including chemical and biological weapons, so these programs lost support. The BW program of the late 1960s in particular was small in size when compared to nuclear, chemical, missile, and conventional arms programs. Further, none of the defense industrial ministries was involved in modern biotechnology; nor were any of the scientific research institutes and scientific production conglomerates. It is reasonable to conclude that the relatively modest cadre dedicated to maintaining the BW program had little influence among those shaping high-level policy decisions on military force–building.

Second, and perhaps more important, applied microbiology for BW purposes, and likely for civilian purposes as well, was unproductive in the 1960s due to the negative influence of agronomist Trofim Lysenko on the biosciences. Beginning in the 1930s, by using doctored data, Lysenko convinced Stalin that when an agricultural technique he had developed, termed *vermalization*, was used on a large scale, it would double or triple the crop yields of Soviet agriculture.[72] Since Nikita Khrushchev, Stalin's successor in 1953, also was a Lysenko adherent, Lysenko held sway until Khrushchev was removed as First Secretary in 1964. But until that year, Stalin and Khrushchev both made political decisions that gave full state support to vermalization and forbade the application of other practices by Soviet agriculture. Even worse, during what came to be called a "lost generation of genetics," vermalization was the only theory that was permitted to be taught and practiced by agricultural scientists, as well as by biologists generally, which meant that rational theories such as the Mendelian theory of inheritance and the Darwinian theory of natural selection were banned from Soviet science. In effect, Lysenko was responsible for convincing Stalin and Khrushchev to suppress the correct bioscientific underpinnings of genetics. Lysenkoism was repressed by the Soviet government after 1964, but its influence continued to some extent in the form of scientists who through seniority maintained leadership positions in laboratories and institutes. For these reasons, in the early 1970s most Soviet microbiologists did not understand molecular biology and, even less so, how to apply genetic engineering for either basic research or bioindustrial applications.

Well-founded genetics and microbiology did not completely disappear during the Lysenko era. There were brave directors of institutes dedicated to chemistry and physics who were strongly anti-Lysenko and who would hire bioscientists, including geneticists, to work in their laboratories under false pretenses. After Lysenkoism was discarded, the bioscientists who had labored in secret were instrumental in establishing institutes that became leaders in biotechnology R&D. For example, the Institute of Molecular Genetics, which became a leading biological research center, was an offshoot of the Kurchatov Institute of Atomic Energy.[73]

The 1972 Decision to Acquire a Second-Generation Biological Warfare Program

As discussed in detail by Leitenberg and Zilinskas,[74] two individuals were responsible for promoting the establishment of the Soviet Union's "new" offensive second-generation BW program: the exceedingly bright bioscientist academician Yuri A. Ovchinnikov and Colonel-General Yefim I. Smirnov.

The Key Role of Ovchinnikov

Yuri A. Ovchinnikov was born in Moscow in 1934 and graduated in 1957 from the most prestigious university in the USSR, the Moscow State University, with a candidate degree in chemistry.[75] In 1960, he secured a research position at the Institute of Chemistry of Natural Compounds, which had been founded a year earlier by one of the Soviet Union's most famous chemists, Mikhail M. Shemyakin. The institute investigated compounds derived from living beings, such as antibiotics, peptides, toxins, and vitamins.

As Ovchinnikov had extensive contacts in the West, it is likely that he learned from them about the revolutionary developments that led to the discovery of recombinant DNA in the late 1960s and its spinoff technology of genetic engineering in early 1970s.[76] One of Ovchinnikov's colleagues at the USSR-AS was the prominent Soviet bioscientist Aleksandr A. Baev. In 1972, Baev was instrumental in establishing the first Soviet laboratory of molecular biology and the genetics of microorganisms at the Institute of Biochemistry and Physiology of Microorganisms in Pushchino. Baev's description of how he came to recognize the importance of the new developments in the West was probably similar to Ovchinnikov's experience:

> *Scientific events were continuing to develop, however, and my period of genetic engineering began. The works of P. Berg, S.N. Cohen, and H.W. Boyer (1972–1973) heralded the beginning of the era of*

> *recombinant DNAs. Even before this, however, my attention was drawn to J. Beckwith's publication in Nature (vol. 224, p. 768, November 22, 1969) on the isolation of lactose operon. I was similarly affected by the news that the Congress of the USA had granted 10 million dollars from the 1971 budget to support genetic scientists, represented by J. Lederberg. At that time I had already sensed that there were more important events on the horizon in biology, and I began to prepare my research into molecular biology, starting with prokaryotes.*[77]

On the basis of his excellent publication record and the high esteem with which he was regarded by colleagues, Ovchinnikov was elected as a corresponding member of the USSR-AS in 1968 and just 2 years later was elected full academician of the Division of Biochemistry, Biophysics, and Chemistry of Biologically Active Compositions.[78,79]

While most Soviet scientists believed it was a waste of time for them to participate in Communist Party of the Soviet Union (CPSU) activities, some realized that one had to be an enthusiastic party member in order to move up in the management sector of the science establishment. With very few exceptions, all directors and deputy directors of research institutes were members of the CPSU. Ovchinnikov must have understood this while young, because in parallel to his scientific career, he made a political career. Joining the CPSU in 1962, at the age of 28, he advanced rapidly, becoming a candidate member of the CCCP and a member of the Presidium of the Supreme Soviet of the Russian Republic of the Soviet Union in 1973. Without doubt, Ovchinnikov toed the party line, as evidenced by a statement he made in 1985:

> *The Communist Party considers scientific-technical progress as a key factor in the acceleration of the Soviet Union's socio-economic development. This was clearly and convincingly underlined by the resolution adopted by the Plenum of the Central Committee of the Communist Party of the Soviet Union in April of this year. These*

> *resolutions made Soviet science responsible for a critical task of historical significance. Through these resolutions, Soviet science was instructed to concentrate its forces on the most important of courses, to mobilize its entire creative potential to the greatest extent possible, and to support the Soviet Union in its entry into key areas of science and technology. Naturally, this encompasses the very new and rapidly developing field of biotechnology.*[80]

To understand Ovchinnikov's success in persuading top civilian and military decisionmakers that genetic engineering was a key enabling technology, it is necessary to be aware of the relationship between the Soviet state and Soviet science. Nikolai Krementsov described the relationship in his 1996 book, *Stalinist Science*:

> *The key feature of Stalinist science was the total dependence of science on its sole patron, the party-state bureaucracy. . . . Thus, the state apparatus and the scientific community each strove to acquire what it most wanted from the other. The state provided scientists with funds, resources, and great public prestige; the scientific community gave the state expertise and legitimacy in industry, agriculture, and medicine. Each developed various tactics to deal with its partner. The state established strict administrative control over institutional structures, scientific personnel, research directions, and scholarly communications. For their part, scientists cultivated patrons among the higher party-state bureaucrats and skillfully played upon their constantly changing policies and objectives.*
>
> *Although the Soviet scientific community and the state control apparatus have often been treated as separate entities, the actual boundaries between them were frequently blurred. Their symbiosis resulted in their institutional integration and individual co-option. At their apex, the control apparatus and the scientific community were blended and overlapping. Not only did scientists occupy key positions within various state agencies, but also some scientific institutions,*

> *such as the presidiums of Soviet academies, were in fact key elements of the party-state control apparatus itself. Moreover, all appointments to top positions in the scientific hierarchy had to be approved by the highest party officials. In such circumstances, it is hardly surprising that the development of the various Soviet scientific disciplines was greatly influenced by the personal relations between particular disciplinary spokespersons and their powerful party patrons.*[81]

Krementsov also succinctly explains how science was controlled in the Soviet Union by the CCCP:

> *The main instrument of party personnel policy in general was the system of nomenklatura. Nomenklatura was, literally, a list of posts that could be occupied or vacated only with permission from the appropriate party committee. All party committees, from the Central Committee to the smallest one in the countryside, established personnel departments, whose main function was to approve candidates for appointment to any post included in their own nomenklatura. Initially devised for the personnel of party organs and agencies, the system was expanded in the early 1930s into the scientific community.*
>
> *The nomenklatura system was strictly hierarchical—the higher the post, the higher the party committee controlling its personnel. The posts of president, vice-president, and scientific secretary of such central institutions as the USSR Academy of Sciences and VASKhNIL* [V.I. Lenin All-Union Academy of Agricultural Sciences] *were in the nomenklatura of the Politburo. The posts of institute director and editor-in-chief of a journal were in the nomenklatura of the Central Committee Secretariat. The position of laboratory head belonged to the nomenklatura of the regional party committee. Even the post of librarian in a scientific institute was in the nomenklatura of the local party committee. . . . Thus, to occupy any administrative post in a scientific institution, a scientist had to obtain permission*

> *from the party apparatus. . . . Nomenklatura thus became the main means of party control over the scientific community.*[82]

While some aspects of the complex relationship between the party-state and the scientific community changed in the post-Stalin era—for example, some *nomenklatura* positions in scientific institutions no longer absolutely demanded that the candidate be a member of the CPSU—in the Brezhnev era the system largely functioned as Krementsov describes it. This background allows certain conclusions to be drawn about Ovchinnikov's influence within the USSR-AS and government.

In 1967, Ovchinnikov was appointed director of the protein chemistry laboratory at the Institute of Chemistry of Natural Compounds. This position most likely belonged to the nomenklatura of the Moscow regional or city party committee, one of the most powerful committees in the USSR. Before approving Ovchinnikov for this position, committee members undoubtedly learned about him and liked what they saw. A year later, Ovchinnikov was elected as a corresponding member of the USSR-AS. His selection was probably more about academy business than that of the party; nevertheless, such an election would have been unlikely without approval by Moscow CPSU officials, which would have given Ovchinnikov more visibility among important politicians. Then, in 1970, Ovchinnikov was elected full academician of the USSR-AS and honored with the Lenin Prize.[83] This combination of honors could not have happened without the approval of the Politburo. It is reasonable to believe that before being accorded these honors, Ovchinnikov would have met with Politburo members, including CPSU General Secretary Leonid Brezhnev. According to one of his former graduate students, "Ovchinnikov impressed Brezhnev with his imagination and knowledge."[84]

Because of his prominence in the scientific community, in the early 1970s Ovchinnikov was chosen as one of only two scientific advisors to the Politburo. He also served as a scientific advisor to the Military-Industrial Commission (VPK) of the USSR Council of Ministers.[85] He thus was in a favorable position to explain the importance of modern biotechnology for military and, probably, civilian applications to Brezhnev and other government officials. Aware of the USSR's inferiority in the biosciences and fearing that the already wide gap between Western and Soviet capabilities in this field would grow into a chasm, Ovchinnikov likely concluded that the only way to quickly gain support from decisionmakers for a program that aimed to match Western developments was to promote its military benefits. He could do so by convincingly arguing to civilian and military decisionmakers that the U.S. Department of Defense was likely to apply genetic engineering to create new deadly pathogens for weapons applications.

According to Vladimir A. Pasechnik, the first defector from the Soviet BW program, as part of his campaign Ovchinnikov wrote a memorandum to the CCCP sometime in 1970 or early 1971 on the necessity of applying modern biotechnology to develop biological weapons. He reportedly used as a model a memorandum written in 1938 by a physicist addressed to Stalin on the necessity of acquiring nuclear weapons. That memorandum proposed establishing a large nuclear weapons program to be carried out in secret nuclear cities. While Pasechnik never read Ovchinnikov's memorandum, his friends in the USSR-AS recounted to him the essence of its contents. In particular, Ovchinnikov was said to have stressed the need to solve scientific problems related to BW using new biotechnology techniques and that doing so was vital to national defense. In order for the Soviet Union to undertake the program Ovchinnikov proposed, it would need to make a long-term commitment and back it up with large state resources, in much the same way that the country supported its World War II–era nuclear program. Pasechnik was certain that Ovchinnikov could not have written and submitted his memorandum without first having secured strong support from highly placed academicians, including the President of the USSR-AS, Mstislav V. Keldysh.[86]

Ovchinnikov is credited with having said, "At the Central Committee of the Communist Party, if we offer ten drugs nobody would support us. Nobody would give us money for medicine. But offer one weapon and you'll get full support."[87] Whether this quote is accurate or apocryphal, there is a near-consensus among former BW scientists who were interviewed by Leitenberg and Zilinskas that Ovchinnikov was their "big man"—the most influential person in garnering the support from the Soviet political and military systems that led to the Soviet government's decision sometime during 1969–1971 to establish a new, very large offensive BW program.

Recently, the current director of the institute that Ovchinnikov once directed, the Shemyakin-Ovchinnikov Institute of Bioorganic Chemistry, was interviewed on the status of his institute and took the opportunity to idolize his old boss:

> *It was he who convinced the country's leadership of the need for serious work on the problem of our country's biosecurity and being prepared to ward off all threats—both natural and competitive. . . . Ovchinnikov was a genius at communication, and after he succeeded in convincing party leaders and government at the time of the need for biosecurity programs, the machine was put into motion. Three joint Central Committee of the Communist Party of the Soviet Union and USSR Council of Ministers decrees were approved (in 1973, 1981, and 1985), according to which all the work proceeded. . . . Five institutes of the USSR Academy of Sciences system, including our institute, as well as an entire series of institutes, institutions, and enterprises (including the Ministry of Health, Main Directorate of the Microbiological Industry, Ministry of Agriculture, and others) were chosen as the principal executors of these decrees. It was a deeply echeloned, well-conceived effort with strong material and financial support.*[88]

Several scientists interviewed by Leitenberg and Zilinskas said that Ovchinnikov was not personally interested in BW-related R&D; to him it was merely a means to become more politically powerful and to be able to disperse funding to those he favored in the Soviet scientific establishment. Whatever his ultimate objective, as a result of Ovchinnikov's influence the Soviet government in 1971 designated biotechnology as a field of critical importance in the civilian sphere, as well as in the secret military sphere. In the military sphere it enabled the creation of by far the world's largest and most sophisticated BW program, which operated in violation of international law for more than 20 years. However, as the next section of this paper makes clear, while Ovchinnikov's influence was vital, it had to be directed at the MOD's General Staff, which was the ultimate arbiter of decisions related to military force–building involving advanced technologies. And within the General Staff, the major proponent for BW was Smirnov.

The Geopolitical Context

Decisionmaking in the Soviet era was informed by the continuous analysis of the USSR's standing in relation to other countries, as well as appraisals of the most important factors that defined the Soviet Union's status. In the times that are most relevant to this study, the late 1960s and early 1970s, the term "correlation of forces" was frequently used by civilian and military leaders at all levels of government, as well as by writers whose books and articles dealt with Soviet internal and external politics, to describe these assessments. One of the best Western analysts of the Soviet system, Michael J. Deane, describes the concept as follows:

> *The calculation of the correlation of forces takes into consideration numerous economic, military, political, and international factors. In speaking of the correlation of forces, Soviet spokespersons appear to differentiate two levels of analysis: (1) the general, worldwide level, and (2) a level of individual factors and/or regions. Assessment on the first level is essentially an intuitive process in which the correlation of forces is based upon a "feel" for world events. With respect to individual factors or geographic regions, analyses are made more systematically because of the far fewer elements to be assessed at a given moment. Therefore, whereas the overall correlation of world forces may intuitively show the trend of the international situation, a more specific calculation of individual factors (e.g., economic, military, etc.) or in a separate geographic region is needed in order to assist in the formulation of foreign policy actions.*[89]

From the perspective of the General Staff, the correlation of forces in the early 1970s heavily favored the United States.[90] This period was characterized by Colonel-General Andrian A. Danilevich, Senior Special Assistant to the Chief of the MOD's Main Operational Directorate and one of the most credible authorities on Soviet military strategy, as a time of struggle for strategic superiority, where the Soviets continued to lag in key areas such as the quality of missiles, multiple independently targetable reentry vehicles (MIRV) technology, nuclear command and control, and naval strategic systems.[91] This assessment led to the launch of a rapid development program for intercontinental ballistic missiles. It also led to deception efforts as a way to convince U.S. planners that Soviet capabilities were more advanced than was the case. As an example, successful deception led senior U.S. officials to become very concerned about the Soviets' presumed superiority in chemical weapons. Harold Brown, U.S. Secretary of Defense (1977–1980), asserted that "the Soviet Union was likely to use chemical weapons." He expected the USSR to employ chemical weapons (CW) even if the North Atlantic Treaty Organization (NATO) did not and even in the absence of nuclear exchanges.[92] In a similar vein, as Zbigniew Brzezinski, the National Security Advisor to the President (1977–1980), stated, "the Soviets had significant chemical weapons capabilities and they used CW in exercises. In a serious war, they would probably resort to CW, and they might even employ CW in the absence of nuclear use."[93] In reality, the Soviet military appears to have designated a minor role for chemical weapons in both its tactical and strategic doctrines. As Danilevich noted:

> *Chemical weapons were considered to be a secondary means of armed conflict, since with the advent of nuclear weapons chemical weapons had lost their significance. Although chemical weapons are a means of mass destruction, it is incomparable in its consequences with nuclear weapons. It does not lead to the death of humanity, but it does carry enormously tragic consequences. But they are limited and localized in nature. They were developed primarily as a secondary means in the conduct of armed conflict. Despite the relative unimportance of*

> *chemical weapons, the Soviet Union could not concede to the U.S. superiority in this field and matched all U.S. means, including delivery and agents used. We could deliver it by means of aircraft bombs, and rockets, in sufficient amounts. The arsenals were on the order of 1,000s of tons.*[94] *So we were ready for chemical warfare, but only as a retaliatory means.*[95]

As the Politburo allowed the General Staff to take the lead in decisionmaking regarding military systems, its assessment of the correlation of forces vis-à-vis the United States would have been of the highest importance in informing a decision on whether to initiate a new BW program. Likewise, with China, where a period of turbulent political relations led to a significant buildup of armed forces along the border and a running series of military skirmishes, culminating in the border conflict of 1969, the General Staff's assessment also would have been of the highest importance. While Soviet forces were far superior qualitatively and included tactical nuclear systems, China's willingness to engage these forces appeared to reflect, in the words of a 1969 Central Intelligence Agency memorandum, "a distinctly Maoist method of deterrence. By assuming a hard line posture, Peking was demonstrating to Russia [*sic*] that despite its pre-occupation with internal problems it was determined to resist Soviet pressures and to defend China's territorial rights, while at the same time calling world attention to the Soviet 'threat.'"[96]

A second memorandum in 1970 suggested how Moscow might react to China's hardline stance:

> *Soviet protests over the continuing border clashes contained hints of military action against China; and prominent Soviet leaders, such as party chief Brezhnev and Foreign Minister Gromyko, publicly attacked Mao and his regime. By late summer, Soviet pressure took a more ominous turn. Soviet officials began soliciting reactions to the possibility of Sino-Soviet hostilities, including a Soviet pre-emptive strike against China's atomic installations.*[97]

The significance of these developments is that the Soviets were considering "more severe military measures" against China and informing the world of this possibility. In April 1970, two high Soviet officials, including Kosygin's son-in-law, told Americans that in their opinion "eventually it would be necessary for the USSR to destroy China's nuclear arsenal, even if it meant using nuclear weapons."[98] In June, an *Izvestiya* editor asked an unnamed U.S. official "what the American response would be to a Soviet attack on China."[99] It is probable that when these threats reached the Chinese leaders, it led them to ratchet down their border provocations and agree to bilateral talks in September 1970. After the Soviet Union dissolved, more was learned about the General Staff's thinking during the border conflicts with China in the 1960s. As Danilevich observed:

> *The Soviet MOD was forced to create groupings of forces in the Far East. In the late 1960s and early 1970s the only area that demanded significant force buildup was along the Chinese border. China represented a major diversion of resources and attention. For every one General Staff exercise carried out in the West, three were done in the Far East.*[100]

It must have been difficult for the Soviets to assess the correlation of forces vis-à-vis China in this period. It was obvious that the USSR was superior at the strategic level, given the disparities in the two sides' nuclear capabilities, and in terms of the quality of conventional forces. Nevertheless, a large number of China's huge population could be deployed by Mao along the 4,300-kilometer Sino-Soviet border. This possibility was made vividly clear in the summer of 1966 when an estimated 2 million Chinese converged along the length of the border to demonstrate support of Chinese territorial claims. If a border war was imminent, it would force the Soviets to deploy large numbers of troops to protect against possible intrusions at many sites that could occur at any time. In this kind of situation, the use of nuclear weapons might not be an efficient solution and also ran significant political risks. This uneasy situation would not abate, from the General Staff vantage point, until the late 1970s when it felt increasingly confident that the correlation of forces indeed favored the Soviet Union for several reasons, including China's limited military-industrial capacity and infrastructure to support the projection of power toward the USSR. Moreover, the Chinese did not appear to have any intention to attack the Soviet Far East.[101]

Soviet Distrust of Nixon's 1969 Executive Order

An important factor shaping Soviet decisionmaking was distrust of the U.S. decision in November 1969 to terminate its offensive BW program. Nixon's executive order to this end was implemented during 1970–1972 and the disposal operation that rid the United States of all its BW agents and munitions was witnessed by members of the U.S. and foreign press. However, the Soviets were to all appearances not impressed. In August 1972, a team of four Soviet officials led by Minister of Health Boris Petrovsky visited former U.S. BW facilities that were in the process of being converted to National Cancer Institute laboratories. As *Science* reporter Nicholas Wade noted at a press conference held after the visit, "Petrovsky complained about the 'superficiality' of his visit." Wade further observed:

> *The Russian party saw the building much as the Army had left it, as the conversion to cancer research had hardly begun. But if the Russians were impressed by the significance of the switch* [conversion], *they failed to show it. Maybe they suspect that offensive biological warfare research still continues.*[102]

When Leitenberg and Zilinskas were interviewing former Soviet scientists who had worked in secret BW laboratories, they learned that all of them had to attend Committee for State Security of the Soviet Union (KGB) briefings during which they were told that the United States was conducting offensive BW R&D and therefore the Soviet Union had to do the same. Although some found these briefings to be tendentious and a crude appeal to patriotic duty, military scientists or military planners might have believed what they heard. Their sources of information typically would be TASS (the Telegraph Agency of the Soviet Union), *Pravda*, *Izvestiya*, and technical journals published by government print shops, while classified information was provided to them by the KGB and the Main Intelligence Directorate (GRU). Undoubtedly, highly placed KGB and GRU directors would have had access to raw intelligence and therefore would know that the United States had divested itself of the offensive BW program, but this data would not necessarily have been passed to lower ranked soldiers and civilian decisionmakers. There are many known examples of the GRU providing vastly exaggerated figures on U.S. military capabilities to the General Staff, which in turn would so inform members of the Politburo and thereby secure funding for armament projects favored by the military-industrial complex and their supporting generals.[103] Lacking information to the contrary, it would be difficult or impossible for Soviet decisionmakers to dispute assessments that indicated a continuing U.S. BW program. The scenario used then, as now under the Putin administration, is that U.S. Government laboratories and facilities where BW R&D was conducted were indeed closed or converted, but that subsequently, BW-related projects whose intent violated the Biological Weapons Convention (BWC) cleverly were assigned to academic institutions and biotechnological and chemical industries.

The Military-Technical Rationale and the Decisionmaking Apparatus

As indicated in the figure, the structure that supported force-building decisions as it existed during the 1970s integrated policy input from a variety of sources in the military-industrial and scientific complex.[104] We understand the broad output of this process concerning the large, sophisticated BW program initiated sometime during 1969–1971, but there are many gaps in our knowledge of specific decisionmaking in the biological field, especially when compared to what is known about other top secret programs such as the nuclear, chemical, and missile programs. For example, the study done by Hines et al. is based largely on information on strategic planning and intentions provided by former high-level Soviet planners and analysts. The interviews that generated this information took place during 1990–1993. The author's estimate is that approximately 90 percent of this information deals directly or indirectly with nuclear weapons and their means of delivery. The remainder addresses chemical and conventional weapons and their means of delivery. The words "biology" or "biological" are not found anywhere in

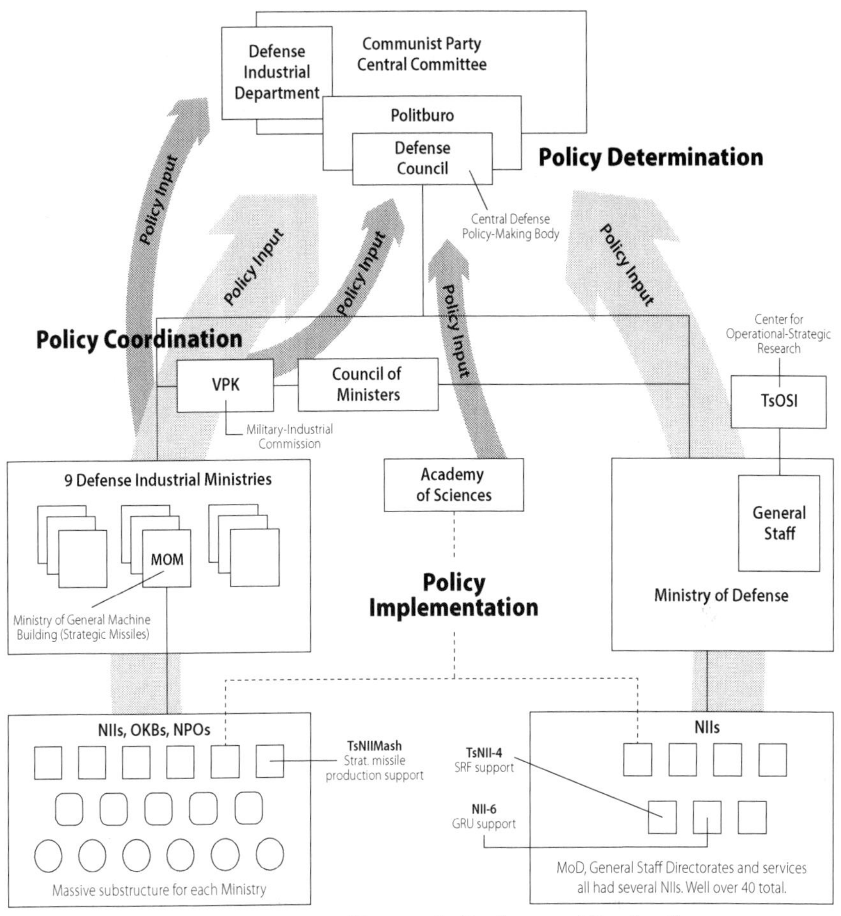

the two volumes containing the results of the study. This is the case despite the fact that Vladimir Pasechnik, the first defector from the Soviet BW program, defected in March 1989 and had undergone extensive debriefings by British and American intelligence officials—a fact that must have been known by some of the individuals interviewed by Hines et al. (although perhaps not by the interviewers). Furthermore, an encompassing book by Andrei A. Kokoshin, one of Russia's best known analysts of Soviet military strategy and thinking, makes no mention of anything biological.[105]

The main reason for this was that all classified programs in the Soviet Union were compartmentalized to an extreme degree, which makes it highly likely that, even within the General Staff from 1969–1971, only a few individuals knew anything about the first-generation BW program, or that its upgrade and expansion were being considered. The BW program had a higher classification level than even the nuclear program because it was about to be deemed illegal under international law, which was not the case with other WMD systems. By signing the Biological and Toxin Weapons Convention in April 1972,[106] the Soviets committed to the global ban on biological and toxin weapons, even though they were already violating its most important provisions. For this reason, the Soviet BW program could not be acknowledged in any way. After the second-generation BW program was instituted, those requiring access to it were accorded F clearance, which was somewhat akin to the U.S. Sensitive Compartmented Information clearance. From 1969–1971, very few civilian officials and military officers were accorded F clearance.

Furthermore, none of the defense industrial ministries was involved in modern biotechnology; nor were any of the scientific research institutes and scientific production conglomerates. It is reasonable to conclude that the relatively modest cadre dedicated to maintaining the BW program had little influence among those shaping high-level policy decisions on military force–building. It was not until Ovchinnikov decided, for reasons described above, to apply his considerable influence to the task, and was able to convince Smirnov of its worth, that the enhanced BW program became reality.

In turn, as the program was based on newly discovered biotechnologies that were largely unknown to the organizations typically involved in force-building decisions, many of them were excluded from the policy determination. This would be true even of key organizations such as the Defense Industry Department of the Communist Party and the Military-Industrial Commission, though the latter came to assume important responsibilities for advancing the second-generation BW program once it was established, so much so that Leitenberg and Zilinskas conclude that its role was crucial to the existence and maintenance of the Soviet BW program.[107]

The Politburo, of course, was the supreme decisionmaking body of the Soviet state. But as Hines et al. report, the Politburo tended to give little attention to military-technical matters. Danilevich told Hines:

> *Brezhnev showed very little interest in the military area and was "very weak" in the area of military decision-making. In exercises he would become very nervous and agitated even thinking about nuclear weapons and would physically tremble when required to make an exercise decision with respect to their use. Because of his aversion to thinking about military questions, he ceded control over military decisions to the Minister of Defense. He also gave carte blanche to the Minister of Defense in terms of defining force requirements. Marshal Grechko, the Minister of Defense until 1976, focused on planning strategic force deployments.*[108]

These observations were seconded by Dr. Vitaly Nikolayevich Tsygichko, a senior analyst at the All-Union Scientific-Technical Institute for Systems Studies, USSR-AS:

> [In] *the Brezhnev era the Politburo delegated all military matters to the Ministry of Defense to include all force procurement decisions. Threat definition was also a military function carried out within the General Staff by the Main Political Directorate. There was essentially no political oversight over the force-building process and no serious challenge from the Politburo to what was clearly a decision situation in which there were serious conflicts of interest. This "hands-off" attitude of the Brezhnev Politburo and the mindless nuclear force-building that resulted was strongly confirmed by General-Colonel Danilevich. . . . Brezhnev and the Politburo left military doctrine to the professionals and gave the military great reign in determining resource allocation and threat definition.*[109]

It seems reasonable to conclude from these observations that a decision on genetic engineering and its possible applications for warfare was not made directly by the Politburo but was delegated to the Soviet MOD and, within it, to the General Staff. As there were important civilian considerations concerning the priority to be accorded modern biotechnology in light of its potential benefits for agriculture, pharmaceuticals, and public health, it is likely that the USSR-AS had powerful influence on Brezhnev and probably other Politburo members. One of Hines et al.'s findings was that "personalities were as important, if not more important, than institutional or bureaucratic competition in determining Soviet military and force-building policy and clearly played a more immediate and decisive role than did expert analysis."[110]

If this was the situation in military force–building, it would have been similar in what one might call civilian force–building that focused on a powerful technology that could have significant positive effects on the Soviet economy. As noted earlier, Ovchinnikov had substantial influence on Brezhnev and, quite likely, the Politburo as a whole, and this would have led to the "there is an opinion" phenomenon. That is, once Brezhnev made known that he was in favor of a course of action as spelled out in an edict, officials serving in agencies that would be implementing that edict would understand that "there is an opinion" and, accordingly, do what was necessary to realize the edict's objectives without questioning it. In this case, once Brezhnev had recognized the importance of modern biotechnology for Russia and made this known, two edicts were drawn up. The first was issued in 1971 and was top secret. It ordered that high priority be given to developing modern biotechnology in the military sphere and that the Biopreparat be established.[111] The direction of military force–building was to be decided by the military. The second edict, issued in 1974, was "On the Measures for Accelerating the Development of Physical-Chemical Biology and Biotechnology and the Use of Their Achievements in Medicine, Agriculture, and Industry," which in effect established a complex program for applying biotechnology throughout the USSR's civilian sphere.

The General Staff was engaged in biological warfare issues through its 7th Directorate, headed since 1954 by Y.I. Smirnov, certainly the most important military decisionmaker in the Soviet BW program.[112] He directed the BW program until

1972, as well as its successor, the 15th Directorate, until his retirement in 1985. Smirnov was one of the principal, and perhaps the most important, of the MOD's BW ideologists, and is also believed to have been the main strategist of biological weapons applications. The Biopreparat scientist Domaradskij called him "the ideologue par excellence of Soviet bioweapons research from the 1950s to the 1980s" and "our apologist for biological warfare."[113]

With respect to advancing the BW program to leverage new scientific developments, Smirnov had two concerns. First, as Domaradskij describes, in the late 1960s he had to confront the faction in the MOD that believed the BW program to be ineffective because it was not "solving problems." These officials also believed there was no need for biological weapons in view of the Soviet's growing strength in the nuclear area. This view might have been strengthened by the U.S. decision in 1969 to close down its offensive BW program because "BW lacked military usefulness."[114]

Smirnov's second concern centered on advances in biotechnology and their possible application by the United States for military purposes as hypothesized by Ovchinnikov. This concern was, oddly enough, first publicly aired by two retired Soviet generals in a 1977 article published in the *Bulletin of the Atomic Scientists*, although its underlying meaning probably was not understood by American readers. The authors noted that R&D in the area of "genetic weapons" had been going on "for a long time" in the United States.[115] The authors provided two specific examples. First, by 1962, the U.S. Department of Defense (DOD) allegedly had confirmed that it was sponsoring research "whose solution would permit discovery of a mechanism which determines the fundamental changes of bacterial cells." Second, in the 1960s, DOD supported a 5-year plan that was said to have "obtained practical results" in transforming a microorganism that gives rise to plague so as to obtain a new strain of this pathogen that was "resistant to antibiotics and does not require a complex nutrient medium for growth."[116] The authors said that they had obtained this information from U.S. Department of Defense Appropriations documents from 1963 and 1970, respectively.[117]

Even earlier than the 1977 article, in 1970 a journal for command-level military personnel published by the U.S. Army Command and General Staff College featured as its lead article "Ethnic Weapons," which outlined the history, desirability, and possibilities of engineered biological pathogens that would affect only those races which historically have no natural defense against certain "enzyme inhibitors."[118] Its author, a medical doctor, explains that many of the chemical activities and functions within the human body are caused by the interactions of enzymes. One of the more significant activities enabled by enzyme chemical reactions is the contraction and relaxation of muscle tissue. If the activities of these enzymes are blocked, the victim will be paralyzed, even to the point of death by asphyxiation.

Undoubtedly, the two articles had been passed to Smirnov and, given his medical background and, presumably, Ovchinnikov's expert explanation of genetic engineering's revolutionary applications, he would have understood their implications for biological weaponry. As is made clear by the interviews conducted by Hines et al., the Soviet military was keen on learning about and acquiring new American technologies. This attitude was rooted in an enduring belief in American technological and scientific omnipotence. In addition, the military is likely to have been affected by a type of mirror imaging; were a similar publication to appear in the USSR, it almost always would signal official approval of the idea or development that is being described.[119]

Being an advocate for retaining and, likely, expanding the Soviet BW program, Smirnov probably acted to exploit this development. He would have sought to make clear that while the U.S. Government had publicized closing its offensive BW program in 1969, in reality it had lied and simply transferred the program from the Defense Department to private companies and university laboratories charged with developing new biological weapon agents.[120] As noted earlier, the GRU would have been helpful in this regard given its practice of routinely exaggerating U.S. threats to the USSR. Confronted with intelligence that the United States was applying revolutionary advances in biotechnology to produce powerful new weapons—information that, however false, would be difficult to dispute—senior General Staff officers who were against continuing the BW program could more easily be persuaded to change their views.

It is possible that Brezhnev may have welcomed the Ovchinnikov and Smirnov initiative to apply genetic engineering in an expanded and improved BW program for internal political reasons in light of improving relations with the

United States. Since 1969, Moscow and Washington had been working toward the "détente" that after the Nixon-Brezhnev summit of May 1972 came to characterize a period of reduced tensions and greater cooperation. In response to internal opposition to détente, Brezhnev is quoted as telling a 1971 Politburo meeting, "We communists have to string along with the capitalists for a while. We need their credits, their agriculture, and their technology. But we are going to continue massive military programs and by the middle of the 1980s will be in a position to return to a much more aggressive foreign policy designed to gain the upper hand in our relations with the West."[121]

The newly empowered BW program would have fit well into this strategy as a major military program that would employ "their technology" and thus serve as a concrete demonstration of Brezhnev's commitment to gain military superiority over the West. It was also realistic to foresee demonstrable military gains from the new BW program by the "middle of the 1980s." Such thinking would have been consistent with the broader and highly urgent imperative in Soviet military planning to keep pace with and respond to American advances in military technology. This issue comes up repeatedly in the Hines et al. volumes. Interviewees peppered their comments with statements such as the following:

- "Qualitative technological advances and R&D efforts were largely conditioned by competition with the U.S. and, in the eyes of the Soviets, were reactive and imitative in most instances."[122]

- "The Soviet military leadership was particularly intent on responding to technological advances in U.S. weaponry."[123]

- "The military wanted a mix and quantity of weapons that supported the General Staff's operational strategy, weapons that embodied the most advanced technologies with which to counter a technologically advanced Western enemy."[124]

- "There is evidence that [in the early 1960s] the VPK and the Central Committee's Defense Department as a matter of policy stressed the need of the Soviet Union copying of foreign technologies and systems, rather than supporting domestic R&D. Almost all sources stated that during the period in question, new systems were developed primarily in reaction to developments in the U.S."[125]

◆ "The Soviets followed the U.S. lead in many technological areas, including MIRVs, missile accuracy, early warning systems and command and control, neutron weapons, low frequency, enhanced EMP [electromagnetic pulse], and *other exotic weapons* [emphasis added]."[126]

Thus the Soviet military seemed predisposed to acquire and adapt for its own purposes almost any new Western technology. This propensity would have been even stronger concerning modern biotechnology because of a convergence of several factors: possibly the brightest Soviet bioscientists had made clear that genetic engineering could be used to develop new, powerful pathogens for military purposes; a top general highly regarded for his knowledge of military medicine favored instituting a second-generation BW program; articles published in the West by eminent authors strongly indicated that the United States was pursuing bioweapons R&D, something GRU intelligence served to confirm; and senior leaders were probably hearing about the huge potential of genetic engineering from prominent scientists and academicians. It was in this context that the General Staff decided to eliminate the 7th Directorate, which was in charge of the first-generation BW program, and establish the 15th Directorate for the purpose of directing the new second-generation BW program. Of course, the MOD has never announced or published its reasons for making this change, but it could reflect the fact, as noted above, that the first-generation BW program had by the end of the 1960s stagnated and become unproductive. Perhaps it was viewed as necessary to create a new organization in order to underscore the importance of and give greater impetus to the new BW program.

There probably were discussions in the General Staff on how biological weapons might be useful in the Soviet force structure. If a strong argument could be made that acquiring and deploying advanced biological weapons would substantially strengthen the Soviet military, it would be easier to convince doubters, if any, to go along with establishing the second-generation BW program. When interviewing former Soviet BW scientists, Leitenberg and Zilinskas always posed questions on how biological weapons were to be used. As noted above, none of the interviewees possessed the high clearance level required to have had access to the information needed to provide concrete answers. Nevertheless, they had opinions based on information and ideas they received from military colleagues that allowed them to provide educated conjectures.[127] According to these scientists, the General Staff conceived of five types of war:

- an instantaneous, full-scale nuclear war in which strategic nuclear forces play the principal role
- a protracted nuclear war involving all the armed forces
- a major war in one or several theaters of military operations involving the restricted use of nuclear weapons
- a major conventional war
- a local war involving conventional weapons.[128]

Any of these would be characterized by three levels of military operations—tactical, operational, and strategic, although these could overlap. Tactical refers to engagements between small units at the local level. Operational refers to the enemy's rear areas containing vital facilities such as harbors, airports, supply depots, and assembly areas. Strategic refers to wide-ranging military actions designed to win wars. Since the Soviets did not develop tactical biological weapons, only biological weapons designed to be used at the operational and strategic levels need to be considered. It is not obvious what a protocol for the use of biological weapons in any of the five types of war would look like or at which level of warfare they would be used and under what circumstances. However, some ideas can be developed using the correlation-of-forces discussions presented above, especially those that focused on the United States and China.

The General Staff's greatest concern was war with the United States. During the years that are of most direct interest to this paper, 1969–1971, the prevailing view was that if a war were to break out, it would be of the worst kind—an instantaneous, full-scale nuclear war that would feature massive exchanges of intercontinental missiles, as well as intermediate-range missile exchanges between NATO countries and the Soviet Union. The correlation-of-forces analysis pointing to U.S. strategic superiority suggested that after such a conflict, the Soviet Union would be in far worse condition than the United States in terms of military and economic power, public health, and other measures. If, however, the Soviet Union could effectively deliver deadly and contagious pathogens over a few population centers in North America, it possessed at least a theoretical capability to initiate a pandemic that would kill or seriously sicken a large proportion of Americans who had survived the nuclear holocaust.

Two pathogens that the first-generation BW program had already weaponized and that would be appropriate for this purpose were the variola virus (which causes smallpox) and *Y. pestis* (which causes pneumonic plague). In nature, smallpox has a fatality rate of about 30 percent and pneumonic plague of about 80 percent if untreated. However, there was at the time an effective vaccine against smallpox, and *Y. pestis* is susceptible to antibiotics and, unless protected, its cells die soon after they are released into the open air. To sell a second-generation BW program to generals, scientists like Ovchinnikov could propose to develop variola virus strains that were more deadly than the viral strains found in nature by inserting genes taken from hemorrhagic fever virus and, similarly, to insert genes into *Y. pestis* cells that make the bacterium resistant to common antibiotics. These are only examples of what actually was done by the second-generation BW program to enhance the infective, virulence, and hardiness properties of bacteria and viruses.[129]

In the second type of war, a protracted nuclear war, Warsaw Pact and NATO forces likely would be involved, supporting a similar argument for BW use. That is, biological weapons would be used against survivors of nuclear exchanges. As the Soviet BW program also developed biological weapons to sicken animals and plants, these weapons could also be employed were the war expected to continue for a protracted period. The dispersal, for example, of foot-and-mouth disease or African swine fever virus would cause horrendous damage to surviving American and European livestock, while there are several bacterial and fungal pathogens that could be used to destroy barley, maize, rice, rye, wheat, and other crops. It bears noting that all NATO countries would have been poorly prepared to defend themselves against an onslaught of weapons bearing these pathogens, even if nuclear weapons had not been used.

Were the third type of war to occur—one involving several theaters of military operations but restricted use of nuclear weapons—there would be very good reasons for the Soviet Union to use both chemical weapons for tactical purposes and biological weapons at the operational level. Lethal biological weapons likely would be employed against military strongholds away from urban areas and containing few civilians, so as to reduce the risks of nuclear escalation. Alternatively, incapacitating weapons would be used, for example, to sicken defenders of airfields, ports, storage depots, and the like without killing civilians in nearby urban areas and leaving facilities and stores intact so that they would be of immediate use to advancing Soviet forces.

When Leitenberg and Zilinskas asked Soviet weapons scientists about their views on how biological weapons were to be used, a significant proportion of them answered that China would be the target. In their view, China's huge population advantage would have necessitated the use not only of nuclear weapons, but also biological weapons.[130] Since no ocean separates the two countries, it would be imperative to avoid the use of contagious pathogens, such as variola viruses, which could spread easily to countries bordering China. While the Soviet population was almost completely protected by vaccination against smallpox, the vaccination rate was much lower in Afghanistan, Burma, Nepal, Pakistan, and other nearby countries (the rate for China was not known). Instead of a contagious pathogen, it would make more sense to employ a noncontagious agent, such as Marburg virus, which is highly virulent with a morality rate of over 80 percent and for which there is no vaccine or effective treatment. The second-generation BW program, in fact, weaponized the Marburg virus and developed a method for its mass production.

In wars of the fourth and fifth types, involving only conventional weapons, biological weapons probably would not have been used. This supposition is not certain though. For example, a conventional war with China that went poorly for the Soviets could lead them to use nuclear and/or biological weapons to restore their position and gain the upper hand. A similar scenario could also occur in the context of a conventional war with NATO. Soviet military thinking along these lines likely will not be known with any fidelity until MOD archives are opened to scholars, which is not likely to occur while Putin and leaders like him remain in power.

To summarize the foregoing discussion, one can see the military and technological arguments underpinning a final judgment supporting the establishment of a second-generation BW program: that by possessing powerful biological weapons that no other country, especially its major potential adversaries, were aware of and therefore could not defend against, the Soviet Union would gain strategic and operational superiority and thus shift the correlation of forces in its favor. Crucial to this was the ability to inflict a major technological surprise on the United States in nuclear war scenarios, providing the Soviet Union with an advantage in postwar recovery. An additional consideration could have been that the appropriate biological weapon might have been capable of decimating China's population.

What's Past Is Prologue

Understanding the Decision

While the precise deliberations of the Politburo are unknown, it is well known by now that in the end the Politburo decided to expand and update its biotechnology programs in the military sphere, utilizing modern biotechnology techniques, in particular genetic engineering, to create unique strains of bacteria and viruses for weapons purposes. A careful reading of available sources indicates six factors underpinning this decision:

1. The Soviet Union's first-generation BW program, dating to 1928, provided the necessary foundation for advocating and institutionalizing the second-generation BW program. Even if this program was viewed by some military leaders as ineffective, it served as an institutional-bureaucratic base from which champions of adopting genetic engineering could make their case for instituting an enhanced BW program.

2. The large civilian biotechnology industry that used classical applied microbiology techniques to produce pharmaceuticals, dietary supplements, and single-

cell proteins was well-placed to provide a "legend"—or legitimate industrial cover—for a new entity called Biopreparat that ostensibly was a civilian institution but that in fact reported to the 15th Directorate and whose role was to direct the very large civilian component of the second-generation BW program that secretly conducted both offensive and defensive BW R&D.

3. Influential bioscientists in the USSR Academy of Sciences convinced members of the Politburo, including Brezhnev, of the need to import new biotechnology techniques for the purpose of using them to research and develop unique pathogenic bacterial and viral strains that could be used as weapons. This was necessary, they claimed, to respond to similar U.S. efforts. While the bioscientists were not particularly interested in biological weapons, they knew that "playing the military card" was the best tactic to gain Politburo agreement for the large investment required. As biotechnology is dual-use, once the decision was made by the military to acquire modern biotechnology, it inevitably was broadcast to civilian R&D institutes whose scientists were conducting unclassified investigations. The 1974 edict noted above made that possible.

4. The Soviet military was predisposed to acquire new military technologies whenever they were developed in the West and in the Brezhnev era was given ample funding to acquire infrastructure and human capital to exploit them. There probably was no opposition within the General Staff or VPK to the possibility of gaining new, unique weapons via modern biotechnology.

5. The General Staff's military analysis of the correlation of forces reinforced the rationale to acquire the capability to produce modern biological weapons. Such weapons were seen as a potentially decisive factor in the types of nuclear and conventional wars against the United States and China envisioned by the General Staff at that time.

6. There was a high probability that the program could be kept secret. Concealing and denying the program enhanced the prospects of technology surprise directed at adversaries, but also meant foregoing the deterrence value of openly possessing biological weapons. The latter, however, was not an option after Moscow signed the BWC. Successfully keeping its BW program secret allowed Moscow to appear to be in compliance with the treaty. In fact, it was not until 1989,

and more so in 1992, when the Soviet BW program was revealed to the world, that the treaty's weaknesses with respect to verification, investigation of alleged violations, and sanctioning of violators were widely understood.

The Legacy of the Soviet Biological Warfare Program in Russia Today

In 1992, Russia's President Boris Yeltsin admitted that the USSR had operated an offensive bioweapons program in violation of the BWC.[131] He then attempted to close the program but was undermined by the MOD, which kept its biological research institutes operating and their work programs secret. In September of 1992, Russia, the United States, and the United Kingdom signed a trilateral statement by which Russia pledged to allow its Western counterparts access to these bioweapons facilities. But Moscow failed to deliver on this promise, so the military biological institutes remain as secret today as they were in Soviet times. Further, Yeltsin's 1992 admission was reversed by Vladimir Putin after he became acting president in 1999, and Russia's official position since then is that the USSR never had an offensive bioweapons program and had only conducted defensive research as permitted by the BWC.

We cannot know whether new biological techniques, based on genetic manipulation developed since the Soviet BW program supposedly closed in 1992–1993, have been applied by scientists working in the three top-secret MOD biological institutes to create new or improved weaponized strains of bacteria and viruses. In particular, these techniques could be applied by weapons scientists to develop substances that interfere with genes that control behavior or immunological defense systems. Work to this end was at an advanced stage of development at the time of the Soviet Union's dissolution in December 1991 and could have been continued even during Yeltsin's administration, which ended in 1999.

In February 2012, the Moscow newspaper *Rossiiskaya Gazeta* published a long essay authored by then–Prime Minister Vladimir Putin titled "A Smart Defense against New Threats" that included the following passage:

> *What is the future preparing for us? . . . In the more distant future, weapon systems based on new principles (beam, geophysical, wave,*

> *genetic, psychophysical and other technology) will be developed. All this will, in addition to nuclear weapons, provide entirely new instruments for achieving political and strategic goals. Such high-tech weapon systems will be comparable in effect to nuclear weapons but will be more "acceptable" in terms of political and military ideology.*[132]

After Putin was elected president for the second time in March 2012, he chaired a meeting attended by a select group of his ministers. During this obviously staged event, the ministers took turns explaining what their organizations would do to implement 28 tasks that Putin had stated in the February *Rossiiskaya Gazeta* article. Then–Minister of Defense Anatoly Serdyukov stated, "Mr. Putin, we have thoroughly studied your article and prepared a plan for implementing the tasks set there for the Defense Ministry."[133] In particular, Serdyukov promised to implement Task 4—to create weapons systems that use different physical principles.[134] It bears noting that the official Web site of the Russian MOD has an encyclopedia that defines "genetic weapons" as:

> *a type of weapon able to damage the genetic (hereditary) apparatus of people. It is assumed/expected that some viruses can/may serve as the active principle. These viruses are in possession of mutagenic activity (with the capability to cause hereditary changes) and can introduce into a chromosome cells that contain deoxyribonucleic acid (DNA) and even chemical mutations, taken from natural sources by chemical synthesis or biotechnological methods. The primary result of the use of genetic weapons is damage/injury and changes to basic/primary structure of DNA, which can lead to serious diseases and their hereditary transmission.*[135]

In the days directly following Serdyukov's remarks, comments on them appeared on Russian Web sites and in mass media. Some supported the implementation of Task 4 on the grounds that such weapons were being developed by the United States, as they claimed, and that Russia therefore had to do the same.

Table 2. Soviet Second-Generation Biological Warfare Agents

Bacteria	Disease caused by
Bacillus anthracis	Inhalation anthrax
Brucella melitensis (Brucella suis?)	Brucellosis (undulant fever)
Burkholderia mallei	Glanders
Burkholderia pseudomallei	Melioidosis
Coxiella burnetii	Q fever
Francisella tularensis	Tularemia
Legionella pneumophila	Legionnaire's disease
Yersinia pestis	Pneumonic plague
Virus	
African swine fever virus	African swine fever (panzootic)
Capripoxvirus	Goatpox, sheeppox (panzootic)
Foot-and-mouth disease virus	Foot-and-mouth disease (panzootic)
Marburg virus	Marburg virus disease
Rinderpest virus	Rinderpest (panzootic)
Variola major	Smallpox
Venezuelan equine encephalomyelitis virus	Encephalitis

Source: Milton Leitenberg and Raymond A. Zilinskas, *The Soviet Biological Weapons Program: A History* (Cambridge, MA: Harvard University Press, 2012), 79–250.

Foreign reactions differed. For example, articles written by David Hoffman in *Foreign Policy* and Raymond Zilinskas in *Nature Medicine* made clear that it is difficult to conceive of a weapon based on "genetic" principles that would not be a violation of the BWC's core provisions.[136, 137]

There is great cause for concern that well-resourced secret Russian institutes with access to modern microbiology techniques will provide the basis for the Putin administration to establish a third-generation BW program—for two specific reasons. First, cultures of pathogens that were weaponized by the first- and second-generation BW programs presumably are stored in the cell culture collections of secret MOD facilities. These would include genetically engineered strains whose infective and virulence properties have been enhanced to make them more suited for use as weapons than strains recovered from nature (see table 2 for a listing of pathogens weaponized by the second-generation BW program).

Second, it is reasonable to assume that recipes generated by the second-generation BW program that specify how to develop, test, and produce biological weapons are deposited in government archives.[138] These recipes could be adapted to develop payloads consisting of third-generation weaponized pathogens. It is also reasonable to assume that the enhancement of existing second-generation weaponized pathogens through the application of modern genetic techniques is easier than starting from scratch with a nonweaponized microorganism.

At this time, there are no signs that Russia intends to scale back the biological institutes that once directed the Soviet bioweapons program or to destroy related culture collections. In this light, Putin's 2012 statement is ominous. Having been forewarned, it is imperative that the United States and other Biological Weapons Convention state parties try to impress upon the Putin government the need to practice transparency by opening its closed bioweapons institutes to outside access, and to acknowledge the illegal Soviet BW program of the past by revealing its accomplishments—including weaponized pathogens—and then taking the necessary steps to publicly destroy them and the associated recipes. Unfortunately, given its past behavior on bioweapons-related issues, the Russian government is unlikely to take any of these steps. Still, by applying this kind of political pressure, it may be less likely that a third-generation BW program will be realized.

Appendix. Biography of Colonel-General Yefim Ivanovich Smirnov

Yefim Ivanovich Smirnov[139] was born in 1904 and died in 1989. In 1932 he graduated from the Military Medical Academy in Leningrad, and in 1938 from the Frunze Military Political Academy. From 1932 to 1937 he worked as a military doctor. During 1938–1939 he headed the medical service of the Leningrad military district, and from 1939 to 1946 he worked in the military service of the Red Army as head of the Main Military-Medical Directorate on Questions of Organization and Tactics of the Military Service. Thanks to the work of this Directorate, many thousands of lives were saved during World War II and millions were returned to duty.[140] For his important work, he was the only physician in the history of Soviet military medicine to be awarded a general's medal (the Kutuzov Order) for organizing the evacuation of the entire army of wounded from Kharkov City before it was captured by German troops. The uniqueness of this operation lay in the fact that it was accomplished by using empty railroad cars and motor transport returning from the front under almost continuous bombing. When the question of withdrawing troops from Kharkov was discussed at the Supreme High Command headquarters, it was informed that approximately 100,000 wounded soldiers were lodged in the city. The Command called Smirnov, who reported that all the wounded could be evacuated in empty returning vehicles. After this was successfully accomplished, Stalin suggested that Smirnov be awarded the general's medal because "Indeed, he saved an entire army." Stalin ended the meeting at the headquarters by saying that, in his opinion, Smirnov should also receive a Lenin Order for his level-headed initiative. Naturally, there were no objections.

After World War II, on February 17, 1947, Stalin appointed Smirnov Minister of Health. Fortunately, fate proved kind to him during the period of repressions that swept the country in the mid- to late 1940s. Later, it became known that Beria asked Stalin several times for permission to arrest Smirnov after he had been removed from the post of Minister of Health in connection with the affair of the doctors-poisoners in the Kremlin. Stalin ordered Beria to leave Smirnov alone and not to approach him regarding this matter again.

After Stalin's death in March 1953, Smirnov served in several different leadership positions in the Soviet army including head of the 7th Main Directorate of the Ministry of Defense (MOD) and, after it was disbanded, head of the MOD's 15th Main Directorate (codenamed Post Office Box A-1968).[141] Starting in 1975, he was a member of the Interagency Council on Molecular Biology and Genetics under *Glavmikrobioprom* (later renamed the Ministry of Medical and Microbiological Industry). In essence, Smirnov was one of the main Soviet ideologues, if not the main ideologue, in the development of biological weapons and strategies of biological warfare. Naturally, this side of his work was always kept secret.

Smirnov was the head editor of the *Encyclopedic Dictionary of Military Medicine* (1949–1950) and numerous installments of "Experience of the Soviet Military during the Great Patriotic War" (1946–1956). He was a member of the editorial board and the editor of the section on military medicine in the second edition of the *Great Medical Encyclopedia.*

According to his biography, both the Medical and Surgical Society of Canada and the Royal Medical Society of Great Britain elected him as a Distinguished Member. A selection of his works include *Issues of Organization and Tactics of the Medical Service* (Moscow, 1942); *Military Medicine and Nikolai Ivanovich Pirogov* (Leningrad, 1945); *Soviet Military Doctors in the Great Patriotic War* (Moscow, 1945); Special Tasks of the Ministry of Health of the USSR, *Soviet Medicine,* no. 8, 1947; "Military Medicine" in *Forty Years of Soviet Health (1917–1957)* ed. M.D. Kovrigina (Moscow, 1957), 309; *Contemporary Meanings of the Basic Propositions of N.I. Pirogov,* with A.M. Geselevich (Moscow, 1960); "Some Issues of Military Epidemiology," *Military Medical Journal,* no. 12, 1960, and *Military Medical Journal,* no. 1, 1961; *War and Military Medicine, 1939–1945* (Moscow, 1976 and 1979); "Problems of Immunology in the System of Anti-Epidemiological Defense of Military Force in the Great Patriotic War," with N.S. Garin, *Immunology,* no. 3, 1980; *Epidemiological Process (Problems and Resolution)* (Moscow, 1980); *Wars and Epidemiology,* with V.A. Lebedinskiy and N.S. Garin (Moscow, 1988).

Among others, Smirnov's honors include the following: academician of the USSR Academy of Medical Sciences, designated Hero of Socialist Labor, recipient of the Order of Lenin, recipient of the Third Order of Red Medal, and recipient of First Commanding Kutuzov Order.

Notes

[1] Jonathan B. Tucker and Erin R. Mahan, *President Nixon's Decision to Renounce the U.S. Offensive Biological Weapons Program*, Case Study Series 1 (Washington, DC: Center for the Study of Weapons of Mass Destruction, National Defense University, October 2009).

[2] Milton Leitenberg and Raymond A. Zilinskas, *The Soviet Biological Weapons Program: A History* (Cambridge, MA: Harvard University Press, 2012).

[3] John G. Hines, Ellis M. Mishulovich, and John F. Shull, *Soviet Intentions 1965–1985, Volume I: An Analytical Comparison of U.S.-Soviet Assessments During the Cold War* (McLean, VA: BDM Federal, Inc., September 22, 1995); Hines, Mishulovich, and Shull, *Soviet Intentions 1965–1985, Volume II: Soviet Post-Cold War Testimonial Evidence* (McLean, VA: BDM Federal, Inc., September 22, 1995).

[4] A biological weapon has four components: a quantity of pathogens that cause disease, a formulation that protects the pathogens while in storage and after being released into open air, a munition that contains the quantity of formulated pathogens, and a dispersal mechanism that disseminates the formulated pathogens over a targeted population. The Soviet decision to establish Biopreparat and operate *Ferment* and *Ekology* was to conduct research, develop, and test (weaponize) pathogens that arm biological weapons.

[5] V.N. Orlov, ed., *We Defended Russia* [in Russian] (Moscow, 2000). Orlov was a colonel on the staff of the Directorate of the Radiological, Chemical, and Biological Defense Forces Command, Ministry of Defense of the Russian Federation.

[6] Ken Alibek with Stephen Handelman, *Biohazard: The Chilling True Story of the Largest Covert Biological Weapons Program in the World—Told From Inside by the Man Who Ran It* (New York: Random House, 1999).

[7] Alfredo Morabia, "'East Side Story': On Being an Epidemiologist in the Former USSR: An Interview with Marcus Klingberg," *Epidemiology* 17, no. 1 (2006), 115.

[8] Ibid.

[9] Valentin Bojtzov and Erhard Geissler, "Military Biology in the USSR, 1920–45," in *Biological and Toxin Weapons: Research, Development, and Use from the Middle Ages to 1945*, ed. Erhard Geissler and John Ellis van Courtland Moon (Stockholm: Stockholm International Peace Research Institute, 1999), 153–167.

[10] According to H.L. Gilchrist, chemical weapons killed 56,000 and wounded 419,340 Russians in World War I. See H.L. Gilchrist, *A Comparative Study of World War Casualties from Gas and Other Weapons* (Edgewood Arsenal, MD: Chemical Warfare School, 1928).

[11] Alibek with Handelman, 33.

[12] Bojtzov and Geissler, 156–157.

[13] Ibid.

[14] *Bacillus anthracis* (*B. anthracis*) is the causative bacterium of the disease called anthrax.

[15] *Clostridium botulinum* is the bacterium that produces botulinum toxin, which is the most toxic substance known to science.

[16] *Vibrio cholerae* is the bacterium that causes cholera.

[17] *Yersinia pestis* (*Y. pestis*) is the bacterium that causes plague.

[18] Article 61 of the new USSR constitution gave the United State Political Administration the responsibility "to unite the revolutionary efforts of the Union Republics in the struggle against political and economic counter-revolution, espionage, and banditism." See Simon Wolin and Robert M. Slusser, "The Evolution of the Soviet Secret Police," in Wolin and Slusser, eds., *The Soviet Secret Police* (Santa Barbara, CA: F.A. Praeger, 1957), 10.

[19] This person might have been M.M. Faybich, who later was recognized for his work in developing various vaccines.

[20] An *oblast* is an administrative unit akin to a province.

[21] Igor V. Domaradskij and Wendy Orent, *Biowarrior: Inside the Soviet/Russian Biological War Machine* (Amherst, NY: Prometheus Books, 2003). In the era before antibiotics, a person who contracted plague usually died.

[22] Kirov is located approximately 500 kilometers northwest of Moscow.

[23] Bojtzov and Geissler.

[24] Aleksandr I. Solzhenitsyn, *The First Circle*, trans. Max Hayward, Manya Harari, and Michael Glenny (London: Harvill Press, 1968).

[25] Gaysky was the inventor of the Gaysky Live Vaccine, which is a preparation of *Francisella tularensis holarctica*, Strain 15; new versions of this vaccine are still in use in the former Soviet Union.

[26] The "convention" Voroshilov is referring to is the 1925 Geneva Protocol, which forbids the use of bacteria in warfare.

[27] W. Duranty, "Soviet Threatens to Use Gas in War," *The New York Times*, February 23, 1938, quoted in *The Problem of Chemical and Biological Warfare, Volume I: The Rise of CB Weapons* (Stockholm: Stockholm International Peace Research Institute, 1971), 287.

[28] J.W. Barnes, C. Henze, W.J. Cromartie, and J.W. Hofer, *A Review of German Activities in the Field of Biological Warfare*, Intelligence Report No. B-C-H-H/305 (Washington, DC: War Department, ALSOS Mission, 1945).

[29] Possibly Fort Alexander I, which was the home of a plague laboratory before 1917.

[30] Walter Hirsch, *Soviet BW and CW Preparations and Capabilities: The Hirsch Report* (Washington, DC: U.S. Army Chemical Intelligence Branch, May 1951).

[31] Kei'ichi Tsuneishi, *The Germ Warfare Unit That Disappeared: The Kwantung Army's 731st Unit* (Tokyo, 1982); Peter Williams and David Wallace, *Unit 731: The Japanese Army's Secret of Secrets* (London: Hodder & Stoughton, 1989); and Sheldon H. Harris,

Factories of Death: Japanese Biological Warfare 1932–45 and the American Cover Up (London: Routledge, 1994).

[32] Olivier Lepick, "French Activities Related to Biological Warfare, 1919–45," in *Biological and Toxin Weapons*, ed. Geissler and van Courtland Moon, 70–90.

[33] Gradon B. Carter and Graham Pearson, "British Biological Warfare and Biological Defence, 1925–45," in *Biological and Toxin Weapons*, 168–189.

[34] Donald Avery, "Canadian Biological and Toxin Warfare Research, Development and Planning, 1925–45," in *Biological and Toxin Weapons*, 190–214.

[35] John van Courtland Moon, "U.S. Biological Warfare Planning and Preparedness: The Dilemmas of Policy," in *Biological and Toxin Weapons*, 215–254.

[36] Erhard Geissler, "Biological Warfare Activities in Germany 1923–45," in *Biological and Toxin Weapons*, 91–126.

[37] *Biological and Toxin Weapons.*

[38] Ibid.

[39] Y.I. Smirnov should not to be confused with L.V. Smirnov, who was head of the Military-Industrial Commission.

[40] Orlov.

[41] Igor V. Domaradskij, personal communication, 1999.

[42] The term "recipe" refers to the written procedures followed by weapon scientists to research, develop, test, and produce a particular pathogen for weapons use.

[43] Orlov.

[44] E.N. Shlyakhov and E. Rubinstein, "Human Live Anthrax Vaccine in the Former USSR," *Vaccine* 12, no. 8 (1994), 727–730; Orlov.

[45] Orlov.

[46] It was the custom in the USSR, and in other countries, to name an important strain after the institute where it was first developed.

[47] Orlov.

[48] Scarification involves the injection of the vaccine by puncturing the outermost layer of the skin (ectoderm) with a needle, but without drawing more than a minimum of blood. This causes a local infection that stimulates antibody production by the immunological defense system of the recipient of the vaccine.

[49] Orlov.

[50] Shlyakhov and Rubinstein.

[51] V.V. Tomilin and R.V. Berezhnov, "Exposure of the Criminal Activities of the Japanese Militarists in Preparations for the Use of Bacteriological Warfare" [in Russian] (1985).

[52] Union of the Soviet Socialist Republics (USSR), *Materials on the Trial of Former Servicemen of the Japanese Army Charged with Manufacturing and Employing Bacteriological Weapons* (Moscow: Foreign Language Publishing House, 1950).

[53] Ibid., 26.

[54] Ibid., 25–28.

[55] Theodor Rosebury and Elvin A. Kabat, "Bacterial Warfare," *Journal of Immunology* 56 (1947), 7–96; Theodor Rosebury, *Peace or Pestilence: Biological Warfare and How to Avoid It* (New York: McGraw-Hill, 1949).

[56] Rosebury headed the Airborne Pathogen Laboratory at Camp Detrick during World War II. Among other accomplishments, he had directed experiments to test *Serratia marcescens, Bacillus globigii, Brucella suis, Malleomyces mallei*, and *Pasteurella tularensis* for use as biological weapons agents or simulants. Rosebury's work and ideas were important in early U.S. postwar decisionmaking on biological weapons research projects requirements. His writings make clear that several bacterial species could be applied for BW purposes and therefore posed dangers to all of humanity.

[57] Jacquard H. Rothschild, "Germs and Gas: The Weapons Nobody Dares Talk About," *Harper's Magazine* 218, no. 1309 (1959), 29–34; and Jacquard H. Rothschild, *Tomorrow's Weapons: Chemical and Biological* (New York: McGraw-Hill, 1964).

[58] George W. Merck, "Biological Warfare: Report to the Secretary of War by George W. Merck, Special Consultant for Biological Warfare, January 3, 1946," *Military Surgeon*, no. 98 (1946), 237–242.

[59] A.M. Arkhangelsky, A.M. Kamorsky, and I.D. Nuzhdin, *Bacteriological Weapons and How to Defend Against Them* [in Russian] (Moscow, 1967) and L.A. Belikov, *The Bacteriological Weapon and Methods of Protection From It* [in Russian] (Moscow, 1960).

[60] The epithet "Red Army" was officially discontinued in 1944.

[61] "A High Reward" [in Russian], *Voenno-Meditsinskiy Zhurnal*, no. 4 (1978), 80–82.

[62] Quoted in Roza N. Lukina and Yevgeniy P. Lukin, eds., *The 50 Years of the Ministry of Defense's Virology Center Deserves Recognition* [in Russian] (Sergiev Posad: Ves' Sergiev Posad, January 10, 2004), 69.

[63] Burgasov was a military microbiologist who in his younger days had worked on a project to weaponize botulinum neurotoxin. From 1965 through 1986, he held the dual position of Deputy Minister of Health and Chief Sanitary Physician of the USSR, which is a position akin to that of the U.S. Surgeon General.

[64] Lukina and Lukin, 69.

[65] Ibid.

[66] Its current name is Russian Federation Ministry of Defense Center for Military Technical Problems in Antibacteriological Defense under the Scientific Research Institute of Microbiology.

[67] V.P. Nepranov et al., "The Analysis of Biological Terrorism Cases (for the Last 10 Years) and Disinfection Actions on Liquidation of Their Consequences," in *Advanced Disinfectants and Safety Techniques Applied in Pathogen Treatment*, ed. G.N. Lepeshkin,

Proceedings of International Workshop (Kirov: Racem Ltd., 2006), 215–228; available at <www.racem.org/>.

[68] Gulbarshyn Bozheyeva, Yerlan Kunakbayev, and Dastan Yeleukenov, *Former Soviet Biological Weapons Facilities in Kazakhstan: Past, Present and Future*, Occasional Paper No. 1 (Monterey, CA: Center for Nonproliferation Studies (CNS), 1999); available at <http://cns.miis.edu/pubs/opapers/op1/index.htm>.

[69] Jonathan B. Tucker and Raymond A. Zilinskas, eds., *The 1971 Smallpox Epidemic in Aralsk, Kazakhstan, and the Soviet Biological Warfare Program*, Occasional Paper No. 9 (Monterey, CA: CNS, 2002).

[70] Domaradskij, personal communication.

[71] Raymond L. Garthoff, "Polyakov's Run," *Bulletin of the Atomic Scientists* 56, no. 5 (September/October 2000), 38.

[72] Lysenko repudiated Mendelian genetics and instead applied vermalization, which was based on the Lamarkian notion that structural changes in animals and plants brought about by environmental or agricultural forces are transmitted to offspring. This notion fitted Soviet concepts on how society can be changed, so it became state dogma avowed by Stalin and Khrushchev.

[73] "History of the Institute of Molecular Genetics," Russian Academy of Sciences, available at <http://old.img.ras.ru/hist_e.htm>.

[74] Leitenberg and Zilinskas, 52–60.

[75] A candidate degree is equivalent to a Ph.D. in the United States.

[76] Genetic engineering had its origins during the late 1960s in experiments with plasmids, which are small, free-floating rings of DNA found in bacteria. A key discovery, made by Swiss microbiologist Werner Arber in 1968, was restriction enzymes, which are able to cut DNA into fragments during replication. The first experiments to combine DNA molecules from different sources were performed in Paul Berg's laboratory in the early 1970s and published in 1972. He shared the 1980 Nobel Prize in chemistry for this work. This was the birth of genetic engineering.

[77] Aleksandr A. Baev, "The Paths of My Life: Autobiography," *Comprehensive Biochemistry* 38 (1995), 439–479.

[78] Scientists who make significant contributions to science could be honored by being elected to membership in the USSR Academy of Sciences (USSR-AS). There were three types of membership: corresponding members, full members (academicians), and foreign members. Being elected to membership was considered very prestigious. In 1974, the Academy had 237 full members and 439 corresponding members. After 1991, the USSR-AS was converted to the Russian Academy of Sciences (RAN); See Answers.com, Russian Academy of Sciences, 2011, available at <www.answers.com/topic/russian-academy-of-sciences-1>.

[79] By the time of his death from cancer on February 17, 1988, Ovchinnikov was a vice president of the USSR-AS, recipient of the Lenin Prize, named Hero of Socialist Labor, and recipient of the State Prize of the USSR.

[80] [Yuri] A. Ovchinnikov, "Biotechnology in the Forefront of Scientific-Technical Progress" [in Russian], *Kommunist* (July 1985), 20.

[81] Nikolai Krementsov, *Stalinist Science* (1996), 4–5.

[82] Ibid., 40–41.

[83] Ovchinnikov was awarded the anniversary medal: "For exemplary work. In commemoration of the 100th anniversary of the birth of Vladimir Ilyich Lenin."

[84] Anonymous, personal communication, 2001.

[85] Anthony Rimmington, "The Soviet Union's Offensive Program: The Implications for Contemporary Arms Control," in *Biological Warfare and Disarmament: New Problems/New Perspectives*, ed. Susan Wright (Lanham, MD: Rowman and Littlefield, 2002) 103–148.

[86] An indication of Ovchinnikov's influence may be realized from what Joshua Lederberg wrote after having met him in June 1985: "Ovch. told me he had gotten Brezhnev's personal backing to modernize Soviet biology through molecular genetics, [fairly explicitly to get over the Lysenko blight] for its indispensable values for medicine and agriculture." See Joshua Lederberg, "Observations from Lederberg during His Presence at the Joint U.S.-U.S.S.R. CISAC Meeting in Moscow in June 1985 [June 12, 1985]," National Library of Medicine: The Joshua Lederberg Papers.

[87] Mark Williams, "Interview with Serguei Popov," at George Mason University (January 3, 2005). A variation of this quote was written by William Kucewicz, "Lead Scientist in a Scourge Search," *Wall Street Journal*, May 1, 1984: "If we bring the Central Committee of the Communist Party of the Soviet Union vaccines, nobody will pay attention to it. But if we bring a virus, oh, then this will be recognized by all as a great victory."

[88] Interview of Vadim Tikhonovich Ivanov, "On the Threshold of Artificial Life" [in Russian], *Zavtra Online*, September 19, 2013.

[89] Michael J. Deane, *The Soviet Concept of the "Correlation of Forces,"* Stanford Research Institute, Strategic Studies Center, SSC-TN-4383-1, May 1976, iv.

[90] Hines et al., *Vol. I*, 23.

[91] Ibid., 23.

[92] Hines et al., *Vol. II*, 13.

[93] Ibid., 16.

[94] This is a gross underestimate; Russia declared having a stock of 40,000 tons of weapons when it joined the Chemical Weapons Convention, but even that number was much lower than the Soviet Union's stockpile.

[95] Ibid., 34.

[96] Central Intelligence Agency (CIA), "Intelligence Memorandum: Sino-Soviet Border Talks: Problems and Prospects," No. 2618/69, November 10, 1969, 8 (Secret).

[97] CIA, "Intelligence Report: The Evolution of Soviet Policy in the Sino-Soviet Border Dispute," No. ESAU XLV/70, April 28, 1970, 9 (Top Secret).

[98] Ibid., 57.

[99] Ibid. The newspaper *Izvestiya* was in effect the mouthpiece of the Soviet government.

[100] Hines et al., *Vol. II*, 23.

[101] Ibid., 154.

[102] Nicholas Wade, "Russians Reserve Doubts: Is Fort Detrick Really De-tricked?" *Science* 177 (August 11, 1972), 500.

[103] Vitaly Shlykov, "Fatal Mistakes of the U.S. and Soviet Intelligence: Part One," *International Affairs* (Moscow) 42, nos. 5–6 (1996), 159–168.

[104] Hines et al., *Vol. I*, xv.

[105] Andrei A. Kokoshin, *Soviet Strategic Thought, 1917–1991* (Cambridge, MA: MIT Press, 1998).

[106] United Nations Office of Disarmament Affairs, "Convention on the Prohibition of the Development, Production, and Stockpiling of Bacteriological (Biological) and Toxin Weapons and on their Destruction"; available at <http://disarmament.un.org/treaties/t/bwc/text>.

[107] Leitenberg and Zilinskas, 565.

[108] Hines et al., *Vol. II*, 23.

[109] Ibid., 138, 142.

[110] Hines et al., *Vol. I*, 7.

[111] Leitenberg and Zilinskas, 564–571.

[112] The 7th Directorate, and subsequently the 15th Directorate, were responsible for the offensive BW work carried out by the Soviet Union. The 15th Directorate was established in accordance with a June 25, 1973, decision of the CCCP (no. 444-138) and a Soviet Ministry of Defense decree of January 11, 1973 (no. 99).

[113] Domaradskij and Orent, 131 and 134.

[114] "New Technology and Biological Warfare. Testimony by Deputy Assistant Secretary of Defense for Negotiations Policy Douglas J. Feith before the Subcommittee on Oversight and Evaluation of the House Permanent Select Committee on Intelligence, August 8, 1986," *Defense Issues* 1, no. 60 (September 8, 1986), 1.

[115] M.A. Milstein and L.S. Semejko, "U.S. Military R&D Through Soviet Eyes," *Bulletin of the Atomic Scientists* 33 (1977), 33–38.

[116] Ibid., 36.

[117] Ibid., 38.

[118] Carl A. Larson, "Ethnic Weapons," *Military Review* 50, no. 11, November 1970, 3–11.

[119] Mikhail Tsypkin, personal communication, September 2, 2014.

[120] There was a basis for this claim. In a report to Congress, the Department of the Army wrote that substantial support had been provided to the pre-1969 U.S. BW program by non-DOD institutions. Specifically, "288 contracts were placed with 73 educational institutions and 440 contracts were awarded to 181 industrial firms." See *Information for Members of Congress: U.S. Army Activities in the U.S. Biological Warfare (BW) Program* (Washington, DC: Office of the Secretary of the Army, March 8, 1977).

[121] Gus W. Weiss, "Duping the Soviets: The Farewell Dossier," Center for the Study of Intelligence, available at <www.cia.gov/library/center-for-the-study-of-intelligence/kent-csi/docs/v39i5a14p.htm>.

[122] Hines et al., *Vol. I*, 7.

[123] Ibid., 11.

[124] Ibid., 49n224.

[125] Ibid, 65–66.

[126] Hines et al., *Vol. II*, 33.

[127] Leitenberg and Zilinskas, 301–309.

[128] Kokoshin, 126.

[129] See Leitenberg and Zilinskas, 698–710.

[130] Ibid., 207.

[131] Viktor Litovkin, "Yeltsin Bans Work on Bacteriological Weapons. This Means: Work Was Under Way, and We Were Deceived" [in Russian], *Izvestiya*, April 27, 1992.

[132] Original article [in Russian] was reprinted by the following source: Vladimir Putin, "Being Strong: National Security Guarantees for Russia," *Rt.com*, February 20, 2012, available at <http://rt.com/politics/official-word/strong-putin-military-russia-711/>.

[133] "Prime Minister Vladimir Putin Holds a Meeting on the Tasks He Set in His Articles as a Presidential Candidate," transcript, March 22, 2012, Archive of the Official Site of the 2008–2012 Prime Minister of the Russian Federation Vladimir Putin, available at <http://archive.premier.gov.ru/eng/events/news/18490>.

[134] Putin fired Serdyukov on November 5, 2012, for alleged fraud and appointed Moscow Regional Governor Sergei Shoigu to replace him. Unless information is received to the contrary, the 28 tasks are assumed to be in force and guiding the Ministry of Defense's future activities.

[135] Ministry of Defense of the Russian Federation, *Encyclopedia: Weapons on New Physical Principles* [in Russian], n.d., available at <http://encyclopedia.mil.ru/encyclopedia/dictionary/details.htm?id=13770@morfDictionary>.

[136] David Hoffman, "Why Did They Do It?" *Foreign Policy.com*, June 18, 2012, available at <http://hoffman.forgeignpolicy.com/posts/2012/06/17/why_did-they-do_it>.

[137] Raymond A. Zilinskas, "Take Russia to 'Task' on Bioweapons Transparency," *Nature Medicine* 8, no. 6 (June 2012), 14.

[138] Leitenberg and Zilinskas, 282–322.

[139] The material in this biography is abstracted from two sources: (1) a eulogy written by Sergei I. Prigoda and inserted in the book edited by Roza N. Lukina and Yevgeniy P. Lukin, *The 50 Years of the Ministry of Defense's Virology Center Deserves Recognition* [in Russian] (Sergiev Posad: Ves' Sergiev Posad, January 10, 2004), 28–29; and (2) A.M. Shelepov, I.T. Leonov, and H. Veselov, *Marshal of Military Medicine* [in Russian] (Kirov: Kirov Military-Medical Academy, 2008).

[140] Smirnov has described the work of the Medical Corps during World War II in an article he wrote in English; see Yefim Smirnov, "The Medical Corps in Red Army Operations: Its Tasks and Their Fulfillment," *British Medical Journal* 1, no. S19 (February 10, 1945), 4388–4390.

[141] The USSR had many secret institutes and agencies that commonly were called "post office box institutes" because their classified names were P.O. Box numbers. The codename for the 15th Directorate was Post Office Box A-1968.

About the Author

Dr. Raymond A. Zilinskas directs the Chemical and Biological Arms Nonproliferation Program at the Middlebury Institute of International Studies at Monterey, where he is also a Research Professor at the Graduate School of International Policy and Management. He is also a consultant to the National Defense University in Washington, DC.

Previously, Dr. Zilinskas worked for 15 years as a clinical microbiologist in acute care hospitals in the Los Angeles area. However, he became interested in policies affecting science and technology and enrolled at the University of Southern California. His doctoral dissertation addressed policy issues generated by recombinant DNA research, including the applicability of genetic engineering techniques to biological weapons development. After graduating, he worked at the U.S. Office of Technology Assessment, the United Nations Industrial Development Organization, the University of Maryland Biotechnology Institute, and, as an Adjunct Associate Professor, the School of Hygiene and Public Health at The Johns Hopkins University.

In 1993, Dr. Zilinskas was appointed Foster Fellow at the U.S. Arms Control and Disarmament Agency, which seconded him to the United Nations Special Commission, where he participated in two biological warfare inspections in Iraq encompassing 61 biological research and production facilities. In 1999, Dr. Zilinskas secured a position at the Monterey Institute of International Studies, where he concentrates on chemical and biological nonproliferation. In 2003, the U.S. Defense and Threat Reduction Agency appointed him as a consultant to manage research projects being undertaken in three Russian research institutes that once had been dedicated to biological weapons research and development. This provided a basis for a long-term study of the history of the Soviet biological warfare program, whose results can be found in the book *The Soviet Biological Warfare Program: A History,* published by Harvard University Press in 2012. He currently is conducting a study of the Putin administration's views on biodefense and biosecurity.

Center for the Study of Weapons of Mass Destruction

Occasional Paper Series

10

The Future of Weapons of Mass Destruction: Their Nature and Role in 2030

John P. Caves, Jr., and W. Seth Carus

June 2014

9

Proliferation Security Initiative: Origins and Evolution

Susan J. Koch

June 2012

8

Defining "Weapons of Mass Destruction"

(Revised and Updated)

W. Seth Carus

January 2012

7

Countering Weapons of Mass Destruction: Looking Back, Looking Ahead

Paul I. Bernstein, John P. Caves, Jr., and W. Seth Carus

October 2009

For additional information, including requests for publications and instructor's notes, please contact the Center directly at WMDWebmaster@ndu.edu or (202) 685-4234 or visit the Center's Web site at http://wmdcenter.ndu.edu.

Center for the Study of Weapons of Mass Destruction

National Defense University

MR. CHARLES D. LUTES
Director

MR. JOHN P. CAVES, JR.
Deputy Director

Since its inception in 1994, the Center for the Study of Weapons of Mass Destruction (WMD Center) has been at the forefront of research on the implications of weapons of mass destruction for U.S. security. Originally focusing on threats to the military, the WMD Center now also applies its expertise and body of research to the challenges of homeland security. The Center's mandate includes research, education, and outreach. Research focuses on understanding the security challenges posed by WMD and on fashioning effective responses thereto. The Chairman of the Joint Chiefs of Staff has designated the Center as the focal point for WMD education in the joint professional military education system. Education programs, including its courses on countering WMD and consequence management, enhance awareness in the next generation of military and civilian leaders of the WMD threat as it relates to defense and homeland security policy, programs, technology, and operations. As a part of its broad outreach efforts, the WMD Center hosts annual symposia on key issues bringing together leaders and experts from the government and private sectors. Visit the center online at http://wmdcenter.ndu.edu.

Made in the USA
Middletown, DE
18 April 2017